THE
PRACTICAL
ENTOMOLOGIST

Above A shield-bug (*Tectocoris diophthalmus*)

THE
PRACTICAL ENTOMOLOGIST

Rick Imes

SIMON AND SCHUSTER
New York · London · Toronto · Sydney · Tokyo · Singapore

A QUARTO BOOK

Simon and Schuster/Fireside
Simon & Schuster Building
Rockefeller Center
1230 Avenue of the Americas
New York, N.Y. 10020

Designed and produced by Quarto Publishing plc,
The Old Brewery, 6 Blundell Street, London N7 9BH

Senior Editor Sally MacEachern
Editor Diana Brinton
Designer Neville Graham
Illustrator Wayne Ford
Index Connie Tyler
Picture Manager Sarah Risley
Assistant Art Director Chloë Alexander

Art Director Moira Clinch
Publishing Director Janet Slingsby

Typeset by Bookworm Typesetting, Manchester
Manufactured by J Film Process
Singapore Pte Ltd
Printed in Hong Kong by Leefung-Asco Printers Ltd,
Hong Kong

1 2 3 4 5 6 7 8 9 10

**Library of Congress Cataloging-in-Publication
Data**
Imes, Rick.
   The practical entomologist / Rick Imes.
      p.   cm.
   Includes index.
   ISBN 0-671-74696-0 (cloth).–ISBN 0-671-74695-2
   (Fireside : pbk.)
   1. Entomology.   2. Insects.   3. Insects –
   Identification.
   I. Title.
   QL463.I44   1992
   595.7–dc20
                                       91-23905
                                       CIP

DEDICATION

The Practical Entomologist is dedicated to the memory
of Robert Wilson Michler of Easton, Pennsylvania,
1919–91.

CONTENTS

THE BASICS OF ENTOMOLOGY

ENTOMOLOGY IN ACTION

DRAGONFLIES & DAMSELFLIES
ORDER ODONATA

GRASSHOPPERS, KATYDIDS, & CRICKETS
ORDER ORTHOPTERA

THE TRUE BUGS
ORDER HEMIPTERA

APHIDS, CICADAS, & HOPPERS
ORDER HOMOPTERA

BEETLES
ORDER COLEOPTERA

BUTTERFLIES & MOTHS
ORDER LEPIDOPTERA

FLIES
ORDER DIPTERA

ANTS, BEES, WASPS, & KIN
ORDER HYMENOPTERA

MINOR INSECT ORDERS

FOREWORD

Entomology, that branch of zoology that deals with the study of insects, is a science often neglected by amateur naturalists – those whose interest in nature constitutes more of a pastime than a vocation. Most of us relate better to birds and mammals because these warm, furry or fluffy creatures embody traits we deem pleasant, such as parental care, live birth (among mammals), and the melodious songs of many birds. Even wild flowers, with their pleasing colors and agreeable fragrances, fare better than most insects when competing for our favor.

Insects, in contrast, with their six legs, hardened body, bulging compound eyes, and antennae, may at first seem more like miniature aliens from a science fiction movie than fellow earthlings. Some insects sting, some destroy our crops, and a few are parasitic or transmit diseases to people or other animals. Many others, however, are extremely beneficial to us, and insects as a group are vital components of the world's food webs and nutrient cycling. Despite our differences with insects, evolutionary evidence strongly suggests that all life on earth had a common origin and that insects merely constitute a different branch on our own evolutionary tree.

Dominant life forms

In fact, a reasonably strong argument can be made for saying that insects, not mammals, are the dominant life form on earth. Consider that the number of insect species is greater than the number of all other species of organisms combined; beetle species alone outnumber all plant species in the world. Despite their conspicuous scarcity in marine environments, insects inhabit every other conceivable habitat, sometimes in mind-boggling densities. Then consider the remarkable resiliency of insects – their short lifespans and tremendous reproductive capacity result in their being extraordinarily responsive to environmental changes that would stress the populations of other organisms.

The Practical Entomologist is designed to introduce you to a fascinating field of natural history and to dispel any aversions to insects that you may have developed. You will find out about many of the insect orders and families and the differences that set each one apart from the others. Together, we will probe the world of insects, examining some of the unique aspects of their lives. The hands-on activities presented throughout the text provide opportunities for you to get personally involved in entomology and make some of your own observations, rather than just reading about someone else's.

Left The world of insects is rich in brightly colored species. Some colors evolved for defense, others for mimicry or for sexual display. This is an African dragonfly (*Trithemis tirbyi*) in typical posture, waiting predaceously on its perch.

Right The wool-carder bee (*Anthidium manicatum*) cards or braids pieces of moss with its legs and mandibles to form a nest. Here, it is collecting pollen for the nest.

THE BASICS OF ENTOMOLOGY

WHAT ARE INSECTS?

Just what are insects, anyway? Often, any small creature with more than four legs is indiscriminately labeled a "bug," but true bugs represent only one of many different groups of insects. What's more, many of these creepy, crawling critters are not insects at all, but may belong to one of several related but very different groups.

Insects, as it turns out, are characterized by several easily recognized traits that set them apart from any other group of organisms. Like other members of the Phylum Arthropoda (which, literally translated, means "jointed foot"), and unlike mammals, for example, insects possess an external skeleton, or *exoskeleton*, which encases their internal organs, supporting them as our skeleton supports us and protecting them as would a suit of armor on a medieval knight. Unlike other arthropods, their body is divided into three distinct regions – the *head*, *thorax*, and *abdomen*. Insects are the only animals that have three pairs of jointed legs, no more or less, and these six legs are attached to the thorax, the middle region of the body.

Most insects possess two pairs of wings, which are also attached to the thorax; the major exception to this rule are the flies, whose second pair of wings is reduced to tiny vestigial appendages that function as stabilizers in flight. Wings, when present, are a sure indicator that an arthropod belongs to the insect class. However, most ants and a number of more primitive insect groups are normally wingless, so the absence of wings does not by itself mean that the creature in question is not an insect.

Adaptability

It was proposed in the foreword that insects could be considered the dominant form of life on earth. Insects have discovered the basic premise that there is strength in numbers. Their life cycles are quite short, less than one year in most cases, and many have a much shorter span, either by design or through predation. They compensate for this by producing astronomical numbers of offspring; so many, in fact, that were it not for the world's insect-eating animals we would surely be overrun within a very short time.

Short lifespans and high reproductivity arm insects with their greatest advantage – adaptability. It works like this: mutations, those genetic variations resulting in physical, biological, or behavioral changes, occur randomly in every population of organisms. When large numbers of offspring are produced, mutations are therefore relatively frequent, and some invariably enhance an individual's ability to compete for its needs or to adjust to changes in its surroundings. Beneficial mutations afford better odds of reaching sexual maturity and passing on the advantageous trait to future generations. Thus equipped, such "improved" individuals can rapidly replace large segments of their species' population that have been decimated by some disturbance in their surroundings.

Top right Some insects, such as this stick insect from San Cristobal in the Soloman Islands, have elongated bodies. Its twiglike form and nocturnal habits allow the insect to move through vegetation virtually undetected. The surface of its integument is warty and armed with a row of spines.

Ants, being social insects, are very industrious. (**Bottom left**), black ants (*Lasius niger*) in a formicarium caring for pupae and larvae. (**Bottom right**), the colorful red skimmer (*Libellula saturata*), from Mexico, rests on a twig. Prominently displayed are the three body parts: head, abdomen and thorax.

CLOSE RELATIVES

There are several groups of animals that could possibly be confused with insects, and all of these are members of the group known as arthropods. Arthropods compose most of the known animal species, and about 800,000 of the 900,000 or so species of arthropods are insects. The others include crustaceans, spiders, centipedes, and millipedes.

Exoskeletons

All have exoskeletons containing varying amounts of *chitin*, a durable organic compound. It was once thought that the amount of chitin present determined the rigidity of the exoskeleton, but more recent research showed that its hardness is proportional to the protein content of the outer layer, or *cuticle*, and that more chitin is found in the soft inner cuticle. In addition to providing protection against injury, the exoskeleton is very water resistant, which inhibits water loss through evaporation. This major evolutionary adaptation allowed arthropods to colonize dry land while other invertebrates were restricted to aquatic habitats.

For all of its advantages, the exoskeleton of an arthropod is also a hindrance. Its weight limits the maximum size that any arthropod may attain, so none becomes very big and the largest are invariably aquatic, where buoyancy helps offset the greater burden. The non-elastic nature of the exoskeleton's outer cuticle is an obstacle to growth, for in order to attain a larger size, hard-shelled arthropods must first shed, or *molt*, their outer layer, which splits open along a genetically-determined seam. Through this opening emerges the now soft-bodied animal, whose elastic inner cuticle can accommodate growth. Those arthropods that rely upon a very hard exoskeleton for defense are particularly vulnerable at this time and often hide until their growth period is over and their armor has again hardened. Most arthropods molt from four to seven times throughout their life.

Also common to all arthropods are bodies that are segmented to varying degrees, jointed appendages (some of which have differentiated to perform specialized functions), and relatively large and well-developed sensory organs and nervous systems, which enable the animals to respond rapidly to stimuli.

Crustaceans

Named for the Latin term *crusta*, meaning "hard shell," nearly all crustaceans are aquatic, and most live in marine environments, although a few of the most familiar, such as crayfish and water fleas, inhabit freshwater, while others, such as certain species of crab, are to be found in brackish water. Lobsters, fairy shrimp, and barnacles are well-known marine crustaceans; sowbugs, those small armored creatures one finds under rocks or in soil, are among the few terrestrial crustaceans.

The head and thorax of crustaceans are combined into one structure, the *cephalothorax*, which may be covered by a shieldlike *carapace*. Their number of paired appendages is variable, but they have at most only one pair per body segment. Only some of these are "legs," attached to the cephalothorax and used for walking. In some species, the first pair of legs are equipped with large pincers modified for grasping offensively or defensively. Other appendages are variously

Top Like insects, crayfish are also arthropods, but they belong to an entirely different class.

Above The shield-like carapace of this crab covers its cephalothorax and is typical of the hard shell that gives crustaceans their name.

adapted for different functions, such as equilibrium, touch, and taste, chewing, food handling, mating, egg-carrying, swimming, and circulating water over the gills.

Some crustaceans are so unusual that their membership in the Class Crustacea can only be determined in their larval stages by zoologists. The barnacles that tend to encrust any marine surface and the water fleas commonly used in high school biology lab experiments are two such oddballs.

Horseshoe crabs
Though unlikely to be mistaken for any type of insect, these "living fossils" are nonetheless arthropods, and the two groups share some very basic features. Horseshoe crabs, named for the shape of their brown, domed carapace, are marine animals. There are two prominent compound eyes, located atop the carapace, as well as two inconspicuous simple eyes. They have a dorsal abdominal shield edged with short spines, and a bayonetlike tail that, despite its formidable appearance, functions mainly to turn the beast over after it has been

Top Unlike insects, spiders have only two major body segments and four pairs of legs. This perfectly disguised crab spider (*Misumena vatia*) consumes a fly.

Above The many distinct segments and legs of this large millipede (*Sigmoria aberrans*) distinguish it from insects.

flipped upside-down by the surf, lest it remain stranded out of water or succumb to ravenous gulls. Horseshoe crabs have six pairs of jointed appendages on the cephalothorax.

Spiders and their kin
Members of the Class Arachnida (from the Greek term for spider, *arachne*) include spiders, scorpions, ticks, mites, and others. It is this group more than any other that is usually confused with insects. Like crustaceans, the body of an arachnid is divided into a cephalothorax and an abdomen. Arachnids have four pairs of jointed legs, all attached to the cephalothorax, although some, like scorpions, possess a pair of large *pedipalpi*, appendages armed with formidable pincers that may resemble legs but are actually modified mouthparts. They also have one pair of *chelicerae*, mouthparts that, among spiders, each terminate with a fang, at the tip of which is a duct connected to poison glands. Unlike either insects or crustaceans, arachnids have no antennae.

Centipedes and millipedes
The name centipede means "one hundred feet," and centipedes are characterized by having one pair of legs per segment; while few centipedes have exactly one hundred legs, the number is a fair estimate. Their long, flattened, multi-segmented bodies comprise between 15 and 181 segments. The head bears a pair of long antennae, a pair of mandibles for chewing, and two pairs of maxillae for handling food. A pair of poison claws on the first segment behind the head enables a centipede to deliver a painful bite if handled carelessly. Most species live under stones or logs, emerging at night to prey upon earthworms and insects, which they kill with their venomous bite.

The prefix "milli-" means thousand, so does a millipede have one thousand feet? Not really, but one might think so to watch this wormlike creature walk. Each of the 9 to 100 or more abdominal segments sports two pairs of legs, this being the chief difference between millipedes and centipedes. The undulating movement of all these legs as the millipede slowly travels is nothing short of mesmerizing. They avoid light, and live for the most part beneath rocks and rotten logs, scavenging dead plant and animal matter. When threatened, they may roll into a tight ball or a spiral to protect their more vulnerable undersides.

TAXONOMY: ORDER FROM CHAOS

Carolus Linnaeus

Taxonomy is the scientific discipline which puts order into an immensely diverse world and allows scientists to discuss any organism and know with certainty that they are talking about the same species. There are two important divisions of taxonomy. *Classification* is the arrangement of organisms into orderly groups. *Nomenclature* is the process of naming organisms.

Common names are generally used in everyday conversation, but they alone do not positively identify a particular species. Many plants and animals have more than one common name, and are often known by different names in different geographical areas, while the same common name may be assigned to two or more totally different species. Clearly, the potential for confusion is great, with well over 800,000 insect species identified and many more still undiscovered.

Contemporary scientists around the world categorize organisms by means of a classification hierarchy, a system of groupings arranged in order from general to specific relationships. They are, in order of increasing specificity: kingdom, phylum (or division, in the plant kingdom), class, order, family, genus, and species. Each of these is a collective unit composed of one or more groups from the next, and more specific, category. Taking them in reverse order, a genus is a closely-related group of species; a family is an assembly of associated genera; an order is a set of similar families, related orders are combined to form a class, similar classes make up a phylum, and all related phylums constitute a kingdom. The complete classification of a honeybee, for instance, is Kingdom Animalia (animals), Phylum Arthropoda (joint-footed animals), Class Insecta (insects), Order Hymenoptera (bees, ants, and wasps), Family Apidae (bumblebees and honeybees), *Apis mellifera.*

All of the above categories are strictly human concepts, and as such they are subject to differences in interpretation throughout the scientific community, even with such a clear-cut system in place. Among taxonomists, there are the "splitters" and the "lumpers." Splitters are inclined to create many subdivisions among organisms, bas-

ing these upon more minute criteria, while lumpers tend to generalize and recognize fewer categories in the same group of organisms. Taxonomy is an active science, and there are occasional changes among accepted classifications that may confuse anyone who does not keep up with scientific literature. In such a case, a glance at the date of the publications containing the questionable terms will indicate which is likely to be the more recent interpretation.

The binomial system

While classification has always been a fairly simple affair, nomenclature has not. By the beginning of the 18th century, the use of Latin in schools and universities was widespread, and it had become customary to use descriptive Latin phrases to name plants and animals. Later, when books began to be printed in different languages, Latin was retained for the technical descriptions and names of organisms. Since all organisms were grouped into genera, the descriptive phrase began with the name of the genus to which the organism belonged. All mints known at that time, for example, belonged to the genus *Mentha*. The complete name for peppermint was *Mentha floribus capitatus, foliis lanceolatis serratis subpetiolatis,* or "Mint with flowers in a head; leaves lance-shaped, saw-toothed, and with very short petioles." The closely related spearmint was named *Mentha floribus spicatis, foliis oblongis serratis,* which meant "Mint with flowers in a spike; leaves oblong and saw-toothed." Though quite specific, this system was much too cumbersome to be used efficiently.

In 1753, Swedish naturalist Carolus Linnaeus introduced a two-word system of naming organisms. This system quickly replaced the older, clumsier method, and came to be known as the Binomial System of Nomenclature (binomial – two names). According to this, individual species are identified by linking the generic name with another word, frequently an adjective. Occasionally, however, the splitters will create two or more subspecies out of what had been a single species, in which case the subspecies name is tacked on after the genus and species, creating a trinomial (three names). All scientific names are Latin, although some have descriptive Greek roots. The first name is always capitalized, but never the second, and both are always either underlined or italicized. When more than one member of the same genus is being discussed, the first name may be abbreviated, as in *D. melanogaster* for *Drosophila melanogaster*.

Phylogenetic tree of life forms

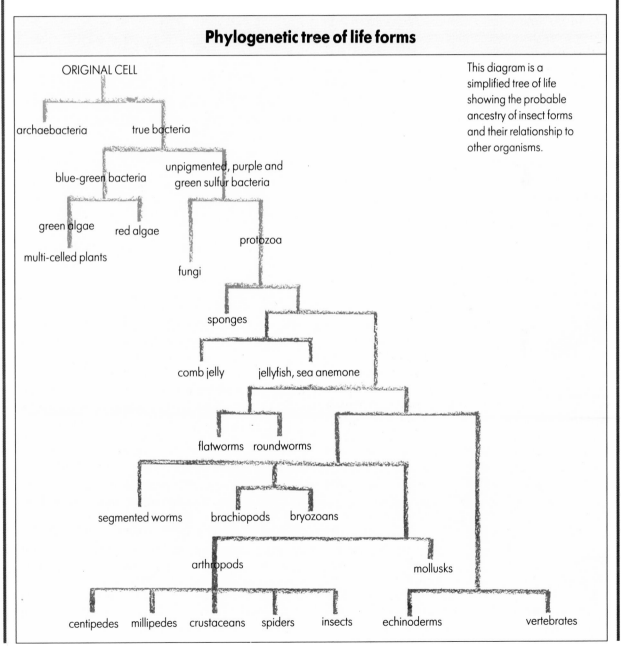

This diagram is a simplified tree of life showing the probable ancestry of insect forms and their relationship to other organisms.

ANATOMY AND MORPHOLOGY

Morphology is the study of external form and structures, the criteria that result in insects being classified as insects and not as something else. Variations on these features define different orders, families, and genera of insects. Related to morphology is anatomy, the internal arrangement of organs and muscles. Learning the basics of both will help you to understand insect lives.

As we mentioned in the beginning of this chapter, the bodies of insects are sheathed in a tough exoskeleton, the hardness of which varies from one species to the next. Because they have no backbone, the support of the exoskeleton is absolutely essential to their mobility on land. The bodies of all insects are divided into three obvious regions – the head, the thorax, and the abdomen.

An insect's head is composed of numerous plates, or *sclerites*, fused together to form a solid capsule that bears one to three simple eyes, two compound eyes, one pair of antennae, and mouthparts. It houses the brain, a fairly simple bundle of nerves from which the nerve cord extends and runs the length of the body along its ventral surface.

The thorax of an insect is divided into three distinct segments. From the head backward, they are the *prothorax*, *mesothorax*, and *metathorax*, each of which is rather box-shaped and composed of four hardened sclerites. The upper (*dorsal*) sclerites of the thorax are called the *notum*, the lower (*ventral*) surface is the *sternum*, and the side (*lateral*) regions are the *pleura* (singular, *pleuron*). Thus, a combination of these terms can isolate any region on the thorax, such as the *pronotum*, *mesosternum*, and so on. A triangular region on the mesonotum, the *scutellum*, is present on all adults, but conspicuous on true bugs (Order Hemiptera).

One pair of legs is attached to each segment of the thorax near the bottom of the pleura. From the thorax outward, the segments of the leg are the *coxa, trochanter, femur, tibia, tarsus*, and *pretarsus*. In addition, adult insects may be wingless or they may have a pair of wings on the mesothorax alone or on both the mesothorax and metathorax.

Below Front and side views of an insect's head showing the plates, or sclerites, forming the solid capsule that bears the eyes, antennae and mouthparts.

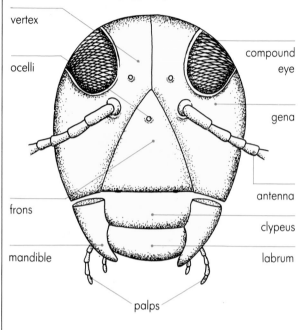

front view

vertex · ocelli · frons · mandible · compound eye · gena · antenna · clypeus · labrum · palps

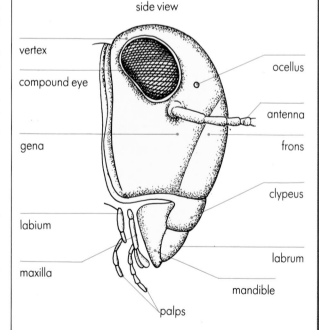

side view

vertex · compound eye · gena · labium · maxilla · ocellus · antenna · frons · clypeus · labrum · mandible · palps

Learning the basics of morphology and anatomy will help you to understand more about how insects work and behave. Study a captured insect and compare its structure with the diagram (**below**) of the external form of a grasshopper. Dissect an insect and look for the mouth parts shown in the diagram (**bottom**).

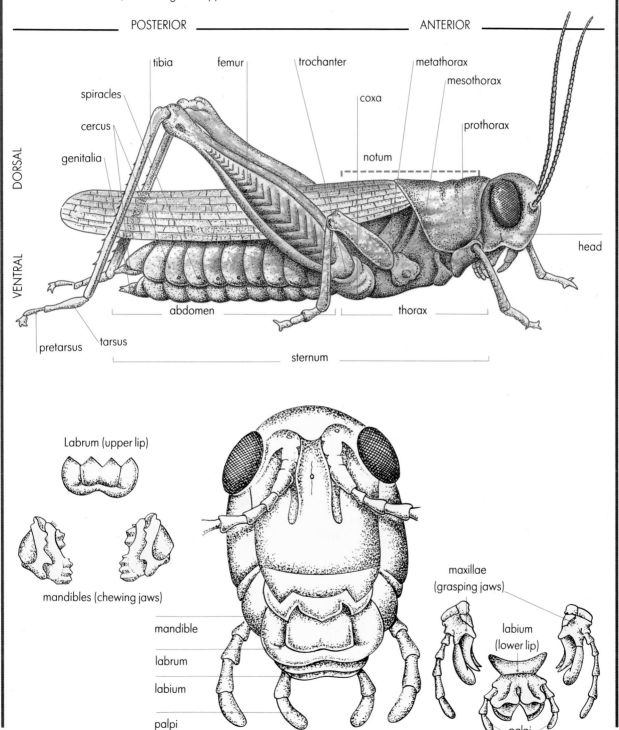

POSTERIOR — ANTERIOR

DORSAL

VENTRAL

tibia

femur

trochanter

metathorax

mesothorax

spiracles

coxa

prothorax

cercus

genitalia

notum

head

pretarsus

tarsus

abdomen

thorax

sternum

Labrum (upper lip)

mandibles (chewing jaws)

mandible

labrum

labium

palpi

maxillae (grasping jaws)

labium (lower lip)

palpi

15

The abdomen

The abdomen, which is softer and more flexible than the head or thorax, consists of eleven segments, although some may be reduced in size and not easily visible. The dorsal surface is called the *tergum*; the ventral side is the *sternum*. It is devoid of appendages except for terminal *cerci* of various sizes and shapes and *genitalia*, or reproductive structures. Females may bear an *ovipositor* for egg-laying, and male genitalia may or may not be extended.

The abdomen is necessarily flexible because it houses the *tracheal system* – the breathing apparatus of the insect – and must expand and contract in order to take in and expel air through *spiracles*, which are openings on each side of the abdomen. There is generally one pair of spiracles per abdominal segment, and they lead to a branching network of air tubes, or *tracheae*, and air sacs throughout the body. From these, oxygen can flow to all organs and tissues, and waste gases can be passed out of the body. Normally, anterior spiracles inhale and posterior spiracles exhale.

Circulatory and digestive systems

Unlike that of vertebrates, the circulatory system of insects is completely independent of their respiratory system and is not involved in oxygen transport. The open system is therefore simple, with a tubelike heart that sucks blood in the posterior end and expels it toward the anterior end. The effect is rather like swirling water in a bathtub, and it makes a stark contrast to our own closed circulatory system in which the blood is always enclosed in vessels, no matter how small. Though inefficient by our standards, insect circulatory systems serve their purpose, since the blood transports only food and waste products.

Despite a number of variations, most insect digestive systems are complete, meaning that a closed tube extends from the mouth all the way through the body to the anal opening, where waste is expelled. There are three main regions: the foregut, midgut, and hindgut, variously modified according to the food eaten by that species.

The strength of insects relative to their small size is legendary; ants, for example, are known to be capable of carrying many times their own weight. These remarkable feats are made possible by the special arrangement of muscles, which are attached to the *inside* of their skeletons, affording tremendous leverage. Muscles are basically attached either within individual segments, enabling the insect to expand or contract, or to adjacent segments, allowing the entire body to flex or simply to curl by coordinating the muscles in a series of segments. Joints generally move only in one plane, so that a series of joints oriented in different planes are necessary to give the legs a full range of motion.

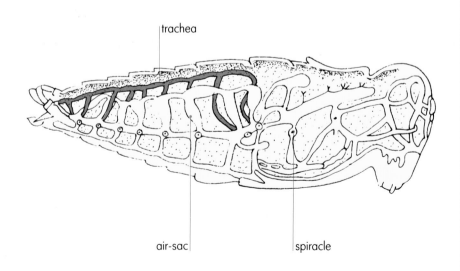

Right The tracheal system of a grasshopper. Air is taken in through the anterior spiracles and circulated through a network of trachea and air-sacs, enabling oxygen to flow to all parts of the body. Waste gases are exhaled through the posterior spiracles.

trachea

air-sac

spiracle

What you can do

Dissecting an insect

You will need:
Equipment (see p45)
dissecting tray
2 probes
forceps
X-acto knife
fine scissors
killing jar
hand lens or binocular/
stereomicroscope
pins

Dissecting an insect will allow you to see first-hand the information presented on these pages. The larger the insect, the better – adult grasshoppers are the best and most easily found, but if you can't find a suitable specimen, the biological supply companies listed in the appendix can supply them for a reasonable price.

1 Sacrifice the insect by placing it in a killing jar (see p54) or in the freezer overnight. Place the thawed grasshopper in a dissecting tray. With the hand lens or stereomicroscope, examine the details of its external anatomy and try to locate all of the features shown in the diagrams. Sketching the whole insect or various parts in a field notebook will help you to remember what you see.

2 Carefully stretch out the wings and notice their appearance and texture (you may use pins to hold the wings in position and free your hands if you wish). The tough, leathery forewings protect the more delicate, membranous hind wings. With the X-acto knife, remove the wings after you have finished examining them.

3 Use a probe to open the mouth and examine the mouthparts – are they used for sucking, chewing, or sponging? With forceps and the knife, separate the mouthparts and try to identify the labrum, mandibles, maxillae, and labium. Which parts have palpi? Which are used for grasping, and which for chewing? (see p.15). Slice a thin layer from the surface of one of the compound eyes and examine it with the hand lens or stereomicrosope. What shape are the facets?

4 Place the grasshopper right side up in the dissecting tray. With the knife or scissors, cut a section from the exoskeleton, being careful not to penetrate too deeply, and remove. Find the abdominal organs shown in the illustration below. Flood the

dissecting tray with water; this will help to separate the organs for a better view, and it will also float the tracheal system, which will appear as a mass of silvery, branched tubes.

5 The heart will be on top if it was not cut away with the exoskeleton. A female may have large yellow ovaries obscuring everything else; push these aside with a probe to find the dark-colored stomach and the Malpighian tubules (a waste removal organ equivalent to our kidneys). Follow the stomach toward the mouth to locate the digestive glands and crop, which is a food storage sac. Along the ventral (lower) wall of the abdomen you may find the nerve cord with its enlarged *ganglia*, from which nerves branch to tissues.

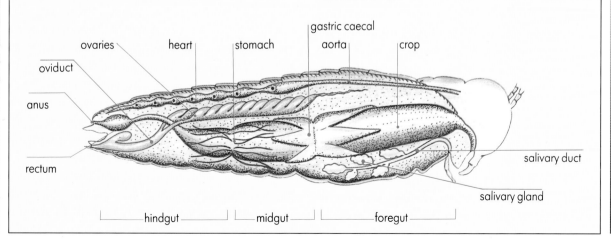

MOUTHPARTS AND FEEDING

An insect's mouth is composed of distinct parts, each serving a specific function. These mouthparts are a clue to the insect's feeding habits and therefore can tell us much about its life cycle and ecological relationships. Among insects, mouthparts are one means of identification, as they are diversely modified to ingest different types of food, but they all fall into one of several categories.

Chewing mouthparts

Chewing mouthparts are the most common type, and are also the mechanism that most closely resembles that of the human mouth. From front to back, chewing mouthparts consist of the *labrum*, analogous to an upper lip; a rather massive pair of toothed, jawlike *mandibles*, adapted for cutting, crushing, and grinding; a pair of *maxillae*, smaller but also jawlike for grasping, and the *labium*, or lower lip. Each of the maxilla is equipped with an antennalike appendage – the *maxillary palps* – which is used for touching and tasting potential food. The labium bears shorter sensory *labial palps*, and is used to guide food into the mouth cavity. Resting on the labium inside the mouth is a tonguelike *hypopharynx*. Major insect groups with chewing mouthparts include dragonflies, damselflies, grasshoppers, crickets, katydids, and beetles. Many bees combine chewing mouthparts with an elongated labium for lapping fluids, especially nectar.

Sucking mouthparts

The other major type of mouthparts are sucking mouthparts. Whereas insects with chewing mouthparts consume solid food for the most part, sucking insects ingest only liquid food, usually plant juices or body fluids. Sucking mouthparts have been modified into a *proboscis*, or beak, composed of an elongated tubelike labium that sheaths the slender, swordlike mandibles and maxillae, which do the actual piercing; these are called *stylets*, and they enclose the food and salivary channels. When you watch an insect, such as a mosquito, about to bite, you can see the labium bend back in the middle to expose the stylets. After the stylets pierce, saliva is injected

Below The mouthparts of an insect help to identify it, reveal its feeding habits, and indicate what kind of life style it leads. These diagrams show some modifications found within the main categories – chewing, sucking and sponging.

Chewing – wasp

mandible

labium

palps

Sponging – house fly

proboscis

maxillary palp

Piercing-sucking – mosquito

maxillary palp

proboscis

Above Typically, apollos are mountain butterflies, laying their eggs on sedums. Here, an adult (*Parnassius apollo*) feeds on knapweed. The proboscis is extended to enable the butterfly to sip nectar.

through the salivary channel, which causes the subsequent irritation of a mosquito bite, then the food is sucked up through the food channel.

The most common examples of sucking mouthparts are the *piercing-sucking* variety, which are found in the true bugs, leafhoppers, treehoppers, fleas, sucking lice, and some flies. *Lacerating-sucking* mouthparts, found on some flies, are similar, but instead of piercing, the stylets are modified to cut the skin minutely, and the fly sucks the blood that flows from the wound. Butterflies and moths do not pierce or cut; they have *siphoning* mouthparts, their proboscis being coiled like a watch spring under the head when not in use and extended to its full length to sip nectar from flowers.

Sponging mouthparts

Most flies have sponging mouthparts that do not quite fit into any of the above categories. A fleshy labium on the proboscis tip acts like a sponge, extending to soak up liquids and food particles.

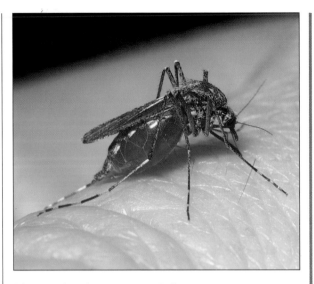

Above A female mosquito (*Culiseta annulata*) fills its abdomen with human blood by means of its syringe-like stylet.

Below Clad in gladiatorial warning colors, especially along its abdomen, this South African grasshopper (*Dictyophorus spumans*) is busy eating a leaf.

What you can do

Predatory games

Equipment:
hand lens
sweep net
examining jar or bug box

If, while collecting, you happen to catch an assassin bug, wheel bug (see p80-7), or any other insect you know to be predatory, you can place it in an examining jar and introduce a likely prey species, such as a fly, in order to witness the piercing-sucking mouthparts in action. A bit of patience is required until both settle down. Of course, the game is rigged in favor of the predator, but you will see nothing that doesn't happen countless times on every acre of land every day during the warmer seasons.

If you're not prepared for such violence, catch a grasshopper and put it in the jar alone. After it settles, put in a crisp piece of lettuce and watch the action of the chewing mouthparts through your hand lens.

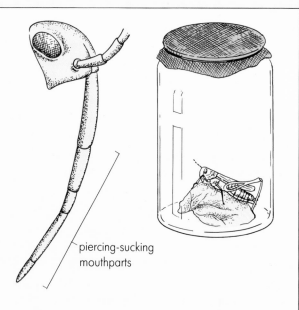

piercing-sucking mouthparts

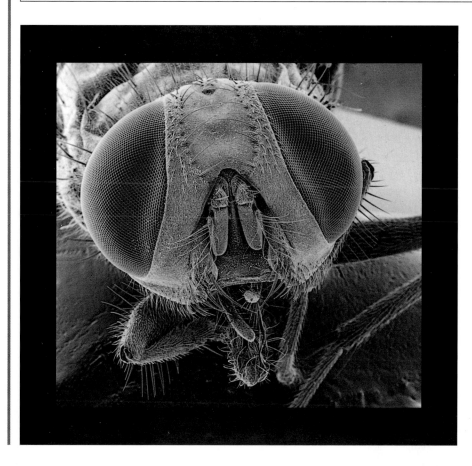

Left The ubiquitous housefly (*Musca domestica*) has been called the most dangerous animal on earth, since it is responsible for transmitting so many diseases to people. One moment a housefly might be on cow dung, the next on sandwiches. Its six feet and sponging mouthparts provide a large surface area with which to transfer pathogens.

21

WINGS AND FLIGHT

The advantages of flight undoubtedly played a large role in the success of the Class Insecta. Insects were the first creatures on earth capable of flight, which allowed them to more easily escape their enemies, to cover more territory in search of food, water, or mates, and to colonize new areas. They could cross large bodies of water, which were insurmountable barriers to most non-flying terrestrial animals.

Except for flies, all flying insects have two pairs of wings, one of which is attached to the upper mesothorax and the other to the upper metathorax. It is likely that their wings originated as flaps that could be extended from the thorax, allowing wingless insects to escape danger by leaping from an elevated perch and gliding some distance away. Insect wings are unique, having evolved specifically for flight, while the wings of birds and bats are merely modifications of pre-existing limbs.

The earliest insects known to be capable of true flight had two pairs of wings that remained extended and did not fold, even when the creature was at rest. Each pair flapped independently of the other pair, a contemporary parallel to this feature being found in the wings of dragonflies, which are members of a primitive but common order. Many advanced insects, such as the beetles, butterflies, and wasps, have evolved means to link their forewings and hind wings together to form two coordinated flight surfaces rather than four.

Most insect wings are laced with distinct *veins*, the pattern of which is often critical to the identification of individual species. The spaces between the veins are *cells*; those extending to the wing margin are *open cells*, and those enclosed by veins on all sides are *closed cells*. Adult insects that emerge from a pupa have wings that at first look crumpled and useless. Extensions of the tracheal system run through the veins, and blood circulates in the spaces around the tracheae. As air is pumped through the veins, the wings unfurl and straighten. As they harden, veins provide both strength and a degree of flexibility, and the wings become capable of sustaining flight.

Veins tend to be thicker and stronger near the body and along the forward, or "leading" edge, and thinner and more flexible near the tip and along the trailing edge. The trailing edge curls on both

Below It is worth becoming familiar with the terms used to describe the areas and venation of butterfly and moth wings as this will help you to identify a particular species when you are consulting a field guide.

Costal (c)
Subcostal (sc)
Radius (r)
Radial sector (rs)

Medial (m)
Cubitus (cu)
Anal veins (av)
Discal cell (dc)

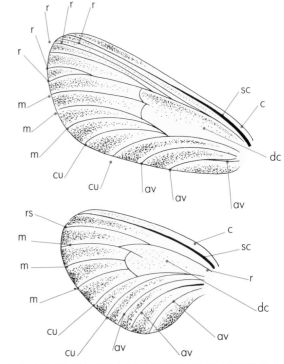

Above Thousands of tiny scales cover the wing membranes of a butterfly, and their varied colors form intricate patterns, as in the yellow bands and eye-spot on the hind wing of this swallowtail (*Heraclides cresphontes*).

Below The wings are lifted as the vertical muscles contract, flattening the thorax (top); then the horizontal muscles contract elongating the thorax (bottom).

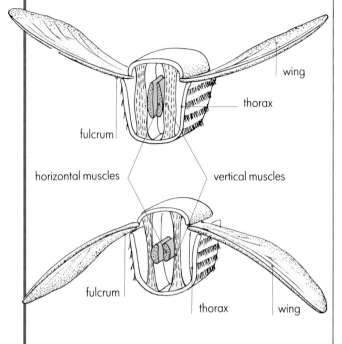

the upstroke and the downstroke, pushing against the air behind it and producing not only lift but forward propulsion and reduced drag.

Propulsion

Insect wings do not move simply by muscles pulling at the base, as one might guess. Instead, two different groups of *indirect flight muscles*, housed inside the thorax, work to alternately elongate and flatten the thorax in a vertical direction. The wings, wedged between the upper and lower thoracic sections, move by leverage on a pivotal point, or *fulcrum*. As the vertical muscles contract, the thorax flattens and the wings move up; then the horizontal muscles contract, pulling the sides in, driving the upper and lower thorax higher, and the wings move down. Smaller *direct flight muscles* at the base of each wing adjust the angle of the stroke and therefore the direction of flight.

The frequency of wing beats varies from species to species, from one individual to another, and even in the same individual at different times. Generally speaking, insects such as butterflies, which have large, light bodies and large wings, need far fewer wing beats to stay aloft than do those with small wings and relatively heavy bodies, such as a housefly or a honeybee. Maximum air speed is also highly variable, but is generally less than 20 miles per hour.

Insects, like most other animals, function more efficiently at warmer temperatures. As cold-blooded creatures, insects cannot rely on their body metabolism to generate heat and must use alternative methods to warm themselves. It is not unusual on a chilly day to see such insects as moths and bees rapidly vibrating their wings as they warm up their flight muscles for takeoff. Others will bask in a patch of warm sunlight; most notable among these are many butterflies that spread their wings and orient the surfaces toward the sun's rays, causing them to function as solar collectors.

Wings have evolved to serve a number of other purposes besides flight. Male crickets and katydids, for example, have developed specialized structures on their forewings; when rubbed together on warm summer nights, these structures produce the pulsing songs by which the males seek to attract females. See pages 38–9 for more about wings.

Above This longhorn beetle (*Adesmus druryi*), taking off from cloud forest vegetation in Costa Rica, displays its brightly-colored first pair of wings, which are modified as hardened and protective elytra.

23

LIFE CYCLES

The vast majority of insects lay eggs, and the development of the embryo progresses outside the mother's body. Most species undergo noticeable changes in form as they mature, a process known as *metamorphosis*. Nearly all insects display either *hemimetabolous* (incomplete) metamorphosis or *holometabolous* (complete) metamorphosis, although a few change so little, except in size, that they are said to have *ametabolous* metamorphosis, meaning that there is practically no change in form.

Incomplete metamorphosis

Hemimetabolous insects are usually distinguished by immature stages, called *nymphs*, that resemble adults, the main changes being an increase in size and the development of sexual organs and wings. Nymphs have mouthparts and compound eyes like their adult forms, and eat the same foods. Their wings begin as external pads on the thorax and develop with successive molts. The stage preceding each molt is known as an *instar*, and each succeeding instar more closely resembles the adult stage than did the previous one. Molting allows for growth, as the newer cuticle is more elastic than the old one.

Among certain hemimetabolous insects, specifically dragonflies, damselflies, mayflies, and stoneflies, the immatures, known as *naiads*, are aquatic and do not resemble their terrestrial adult forms.

Complete metamorphosis

The life cycle of holometabolous insects consists of four distinct stages. From the *egg* hatches a *larva*, whose primary functions are to eat and grow. Wormlike in appearance, larvae do not resemble adults; in fact, they scarcely resemble insects. A larva usually possesses a series of simple eyes on its head, though these may be difficult to distinguish. It will also have either chewing or chewing-sucking mouthparts, a pair of very short antennae, and sometimes three pairs of true legs, although there may be other appendages that resemble legs, or they may have no legs at all. Wings, though developing, are hidden under the cuticle. Larvae molt several times to accommodate growth, and each stage preceding a

Below Nearly all insects lay eggs and the embryo then develops outside the mother's body. There are three types of metamorphosis.

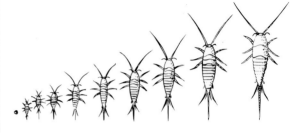

Above A few primitive insects such as silverfish undergo almost no change between the first nymphal stage and the adult form except to increase in size. This is known as ametabolous metamorphosis.

Above Incomplete or hemimetabolous metamorphosis is generally characterized by immature forms called nymphs that resemble adults, or naiads that do not, but gradually increase in size and develop wings and sexual organs.

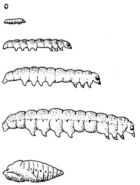

Left More advanced insects undergo complete or holometabolous metamorphosis. The egg hatches into a larva which goes through a pupal stage before an adult emerges that may be completely different in form, habitat and feeding habits.

Above left The tiny 2mm diameter eggs of the puss moth (*Cerura vinula*) are laid in small groups on the undersides of tree leaves, such as sallow, willow or poplar. They hatch in about a week, depending on temperature.

Above right The mature larva is quite unlike the jet black larvae of the early stage. The zig-zag pattern and the nine pairs of white spiracles are clearly visible. When molested, the larva shows off its red head and waggles a pair of long tails to deter predators.

Bottom left The cocoon of the puss moth is not only well camouflaged but is remarkably tough, protecting the pupa. When the larva is preparing to change into a pupa it chews some bark and incorporates this with its silk to make the cocoon.

Bottom right To escape from the hard cocoon, the moth has a special cutting device on its head. The feathery antennae of the male puss moth are unlike the thinner ones of the female. The puss moth gets its name from the furry appearance of its body.

25

molt is known as an *instar*.

At the end of the larval stage, a final molt may occur with a *pupa* emerging, or the last larval "skin" may harden into *puparium*. Pupa do not eat and their movement is usually restricted to no more than a wiggle. In this stage, a great transformation is occurring. Some tissues differentiate, others break down and are reabsorbed and reorganized to form new structures. Through the hardened pupal case, the developing wings can often be seen, as can the compound eyes, antennae, mouthparts, and legs. Inside, reproductive organs develop and the digestive system undergoes modifications.

The pupal stage can last from four days to several months, depending upon the species. At *eclosion*, the *adult* emerges, its wings crumpled and its body soft. Within hours, the wings unfurl, becoming stronger as the veins dry and stiffen, and the exoskeleton also dries, hardens, and gains pigment. In most species, adults have a few weeks to accomplish their primary mission of mating and egg-laying, but their tenure in this stage may last either less than two hours, in the case of certain mayfly species, to several years.

Sexual reproduction

Sexual reproduction among insects is the norm; the male of a species transfers sperm to a female, and the sperm are then stored in a special sac in her abdomen. Here the relationship ends, with each of the pair going their separate ways; in some cases, the male dies soon after mating. Egg-laying, or *oviparous*, females are equipped with abdominal appendages called *ovipositors*, which are variously modified to deposit eggs in a site suitable for their development, always close to an appropriate food source. As the eggs are laid, they meet sperm on the way out of the female. Fertilization occurs through a small opening, the *micropyle*, usually shortly after the eggs are deposited.

Among some insects, the eggs remain inside the mother until they hatch. In that case, if the embryo feeds only on material stored inside the egg, it is *ovoviviparous*. In rare instances an embryo can be *viviparous*, being nourished by the mother's tissues prior to hatching.

Once the eggs are laid, they are abandoned by the mother, who usually dies shortly afterward. As compensation for this and for the often vulnerable

What you can do

Observing metamorphosis

It is said that a leopard can't change its spots, but many insects can. The transformation of an insect from larva to adult is a marvelous spectacle, and one that you can observe in your own home. You will need to make a grasshopper zoo (page 77) or other insect cage (page 51), adapting it for whichever species you choose to raise by including the proper food. This means that you must first be reasonably certain of the identity of your specimen and look up its preferred food in a field guide.

Caterpillars of moths and butterflies are relatively easy to identify and feed. With proper care, you will need to collect only one or two individuals to observe metamorphosis. You can also order specimens and propagation supplies from a biological supply company. The time spent in larval and pupal stages varies from one species to another, but you may also enjoy watching the larvae go through several instars before they pupate. Be sure to release the adults into the appropriate habitat soon after they emerge.

Monarch butterfly caterpillars are easy to obtain, because they are one of the few creatures likely to be found eating milkweed leaves. In fact, you can plan ahead by collecting and planting some milkweed seeds in clay pots and leaving them outside over winter. In spring you will have fresh milkweed plants that you need only enclose in a cylinder made of clear plastic or wire screen with a screened lid. Embed the bottom of the cylinder in the soil of the clay pot, put in your caterpillar, and watch the show!

Above The stridently-colored, bright warning stripes of the caterpillar of the frangipani hawk-moth (*Pseudosphinx tetrio*), from Brazil, defy predators.

nature of the newly-hatched offspring, most species lay a large number of eggs.

Asexual reproduction

Among some insects there exists a type of asexual reproduction – *parthenogenesis* – in which an unfertilized egg will develop into an adult. This is particularly common among social bees and wasps, where the division of labor is drawn strictly along sexual lines. *Workers*, who perform all of the tasks (apart from reproduction) necessary to keep the hive operating, are all sterile females that develop from fertilized eggs, while males, whose sole function is to fertilize the queen, develop from unfertilized eggs.

Below Workers of the yellow meadow ant (*Lasius flavus*) arrange different forms of cocoon in their galleries. The powerful jaws of the workers grasp the loads, searching out a good grip.

What you can do

Egg hunt

You needn't wait until Easter to have an enjoyable egg hunt. Insect eggs are abundant in spring and summer, but even in winter you may find eggs attached to twigs, branches, trunks, or fence posts, or hidden away in cracks. During the warmer months, the eggs of plant eaters are easily found on the surfaces of leaves. Contrary to popular belief, the eggs of many insect species are large enough to be seen if you look in the right places, although a hand lens is always useful for observation. However, if you're expecting typical "egg-shaped" eggs, like those of chickens, you'll be surprised. Insect eggs come in all manner of wonderfully sculpted shapes. Some look like golf balls, others like milk bottles, barrels, spheres, light bulbs, bullets, or flower pots, to name only a few shapes. Eggs may be laid singly, in rows, in clusters, or on top of one another. Those of mantids are encased in a frothy foam that hardens into a tough protective case, and some moths embed theirs in hairs plucked from their abdomens. In addition to the myriad of shapes, insect eggs also come in every conceivable color. Within an individual species, however, all eggs are identical.

Butterfly eggs

Monarch

Silver-bordered fritillary

Falcate orangetip

Tiger swallowtail

INSECT SENSES

We humans tend to imagine that all creatures perceive the world in the same way we do, but such is not the case. Many animals see no colors, while others can see colors we cannot, and vice versa. Some can only detect degrees of light but no images, and still others are totally blind. There are sounds well above or below our range of hearing that are perfectly audible to other organisms, and our sense of smell is notoriously poor compared to that of the so-called "lower" animals. There is much more going on than we realize, and insects know this.

Take color, for instance. Did you know that ultra-violet is a color? We hear about it, and we know that it exists, yet we cannot see it; insects can, however, and they respond to it – a fact not lost on those flowering plants that depend on insects for pollination. Red, on the other hand, a warm and attractive color to us, is invisible to insects.

Eyes
Even the colors mutually visible both to insects and ourselves do not translate into similar images in our respective brains. The head of many mature insects has from one to three simple eyes, or *ocelli*, that serve only to detect various degrees of light. The two *compound eyes* are composed of anything from a few to several thousand individual units called ommatidia, each one evidenced by a separate lens, also known as a facet. Beneath each facet is a second, conical lens, and the two work together to focus light down a light-sensitive structure, the *rhabdome*; this is connected to the optic nerve, which leads to the brain. The quality of the images they generate is not known, but they are supremely adapted to detect motion. Since each unit of a compound eye is stimulated separately, every motion is multiplied many times, and the resulting effect must be rather like watching the same channel on hundreds of television sets at once.

Taste, smell, and touch
The world of insects consists more of patterns of smells and tastes than of light and sound. Taste, smell, touch, and sometimes hearing, are all functions of minute, hairlike bristles called *setae*,

Below A close look at the face of a powerful dragonfly predator (*Aeschna cyanea*) reveals the enormous compound eyes, which meet over the top of the head, giving superb 300° vision.

What you can do

Flower power

Equipment:
assorted photographic filters
about 8½ x 11in of white cardboard
masking tape
flower press
field notebook
pencil
folding table
aerial net
examining jar or bug box

What attracts insects to flowers – scent or color? The answer may be either, depending upon the species of insect in question. You can carry out a simple experiment in your flower garden or in a meadow to determine some of the insects that respond to scent and to various colors.

1 You will need to purchase plastic or glass photographic filters that transmit only one color of the spectrum. If possible, get 7 filters, transmitting only red, yellow, blue, orange, green, violet, and ultra-violet light respectively. You will also need a clear filter as a control. In the garden or meadow, observe which flowers seem to be popular with flying insects. Cut eight blooms of a single species (provided they are not rare or endangered) and press

them in a large book overnight, using tissues or construction paper to blot any moisture. Mount each of the pressed flowers right side up between a filter and white cardboard, and seal the edges with tape.

2 On a sunny day, lay out the pressed flowers on a folding table in the meadow or garden, and observe which ones draw more insects. You may use an aerial net to capture individuals for identification and record them, noting also the type of flower and time of day, in your field notebook if you wish.

3 More than likely, you will observe more visitors at

white cardboard with pressed flower

photographic filter

filter taped over cardboard

the ultra-violet and clear filters than at the others, in which case you may deduce that this particular flower utilizes ultra-violet to attract insects. Insects that visit other flowers, but remain indifferent to the pressed blooms, are probably attracted by scent and not color. Likewise, flowers that

attract few visitors when pressed, but many while alive, undoubtedly rely on their fragrance to invite pollinators. It should prove interesting to repeat this experiment several times, using a different species of flower on each occasion, and also on cloudy days, and compare the results.

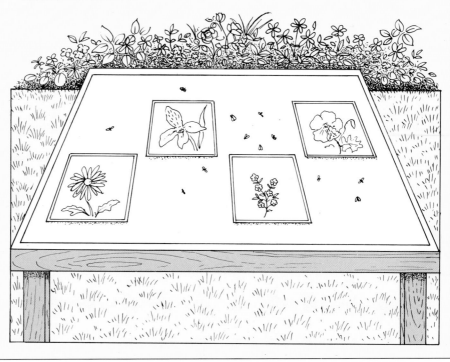

which may occur all over the body but are usually concentrated on the antennae, mouthparts, and legs. Among the *chemosensory* setae, those detecting airborne chemicals account for smell, while those that perceive chemicals through direct contact with solids or liquids allow an insect to taste. *Tactile* setae, which detect touch, are connected to the cuticle by a ball-and-socket joint, the slightest movement of which stimulates nerve endings on the underside of the joint.

Although insects have no noses, their olfactory sense is keener than anything we can imagine. They smell primarily with their antennae, paired appendages located between the compound eyes and above the mouthparts. Segmented, flexible, and densely covered with microscopic setae, antennae have many configurations throughout the insect world. Among their numerous functions, they serve to detect *pheromones* (chemical messages) emitted to communicate with other members of the species, and to locate food, water, and suitable sites for laying eggs. Depend-

ing upon the insect, antennae may also be used to perceive touch, taste, and/or sounds. Insects also taste through the chemosensory setae on their mouthparts and legs.

Sensitivity to sound varies widely among insects, as do the anatomy and location of their auditory organs. *Phonoreceptors*, as they are called, may simply be modified setae on the body, appendages, or antennae that can detect air vibrations, or they may be more complex structures called *tympanic organs*, located on the abdomen, thorax, or the forelegs. A tympanic organ consists of an exposed *tympanic membrane*, covering an underlying air sac. Auditory nerves are connected either to the air sac or directly to the membrane, which vibrates in response to sound waves, stimulating the nerves in a fashion very similar to the human ear. Some insects can differentiate pitch, or various frequencies of sound, while others are tone deaf. As a group, they can detect sound over a much broader range of frequencies than can people.

Right Insects have one pair of segmented, flexible antennae that are densely covered with microscopic setae. Depending on the insects, antennae may be used to taste, touch, smell, detect pheromones, sound, food, water and sites for egg-laying.

Below This exploded view of the microscopic setae which cover the branches of the antennae shows how messages are passed to the central nervous system. The setae are highly sensitive, functioning primarily as organs for taste, touch and smell.

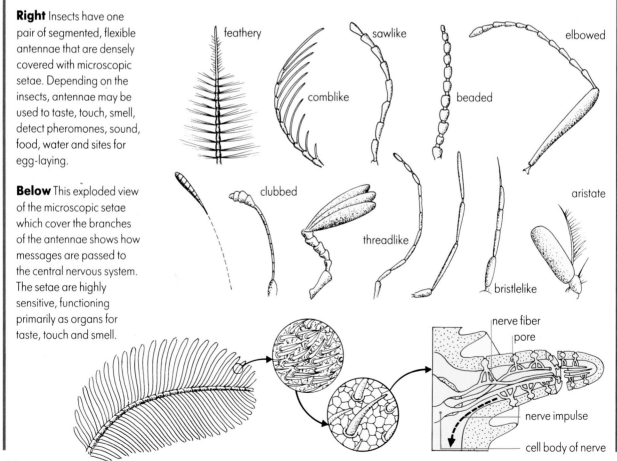

feathery · comblike · sawlike · beaded · elbowed · clubbed · threadlike · bristlelike · aristate

nerve fiber · pore · nerve impulse · cell body of nerve

Above Insect antennae are marvellous sensory detectors. The antennae of males may be more branched than those of females, as in this Indian moon moth (*Actias luna*).

INSECT BEHAVIOR

Behavior is the response of an animal to stimuli from its surroundings. A stimulus is detected through one or more senses and interpreted by the brain, which then instructs the body to react, either on the basis of past experience (learned behavior) or in a genetically pre-determined course of action (instinctive behavior).

Understanding an animal's behavior begins with understanding the motive of its genes, those molecular blueprints of all life forms. Genes are very selfish. They care nothing for the rest of their species, only that their host remains healthy and strong and survives to produce as many offspring as possible, thereby passing on the maximum number of its genes to future generations and ensuring the genes' survival. Even behavior seemingly unrelated to reproduction is ultimately oriented toward keeping the animal fit and helping it survive to sexual maturity.

Insect behavior is largely instinctive. Moreover, the behavior pattern of an insect is practically identical to that of every other member of its species, varying only through random jumblings of a portion of the genetic code, better known as mutations. Mutations occur regularly in all populations, resulting in variations in physical appearance or behavior. Some are beneficial, some detrimental, but most have little effect on an individual's fitness. They are the mechanism of evolution through natural selection.

Mutations

The tremendous success of insects is due to both their stereotyped behavior patterns and relatively frequent mutations. Fixed behavior patterns, guiding insects to react in the ways that best promote their survival and reproductive success, enable them to flourish in their preferred *niche*, that portion of an ecosystem that fulfills their needs. However, the world changes, and the inability to change with it is the chief cause of extinction. Because of the fairly short life cycles and great reproductivity of insects, it is likely that when change occurs there will be mutated individuals floating around within a species that are able to cope better than others, meaning that more of their "altered" genes are passed on. With a high rate of reproduction, and decreased competition

Above This diligent worker honeybee (*Apis mellifera*) is collecting pollen grains from the stamens of a spring crocus, and will carry its load away in the pollen-baskets on its hind legs.

from those that could not cope, it is not long before the beneficial mutant gene displaces its predecessor and the species has evolved to meet the new challenge.

Behavior patterns

It would be impossible to list all of the specific behavior patterns unique to individual insect species. As a practical entomologist, you are best advised to learn the basic behavior patterns of the animal kingdom and to try to recognize these among the insects you observe. Significant behaviors will be discussed later in individual chapters on insect orders. The following are some common categories of behavior.

Courtship is ritualized behavior, employed to attract a mate, and is usually actively engaged in by males. The female response, also highly stereotyped, indicates acceptance or rejection.

Copulation, the physical joining of members of the opposite sex, culminates with the transfer of sperm from the male to the female.

Egg-laying is an important clue to the life cycle of an insect, because individual species are quite stringent in their site requirements. Eggs are nearly always laid either on or with easy access to a food source suitable for the larvae or nymphs.

Feeding may encompass predation, grazing, or scavenging. Social insects actively gather food and return with it to their colony to store it or to share with others.

Defense may be either active, as in insects that sting, bite, or emit noxious chemicals, or it may be passive, employed by those that hide in crevices and under stones, or it can take the form of camouflage in the case of species that "hide" out in the open on a matching background.

Communication, the deliberate transfer of information, occurs essentially between members of the same species. Insects may communicate via all five senses known to us, and we must not discount the possibility of their communicating through unknown methods as well.

Grooming is just as important to insects as to the furry animals to which it is normally attributed. Insects' antennae, eyes, wings, legs, mouthparts, and the hairlike sensory setae covering their bodies must all be kept clean in order to function efficiently.

What you can do

Insect semaphores

Equipment:
penlight
aerial net
examining jar or bug box

One of the favorite summer activities of rural children in temperate regions is to catch fireflies outside on a balmy evening. You can take this fun a step further by learning the language and "talking" with fireflies.

Firefly beetles have luminous chemicals in their tip that can be turned on and off as easily as we operate a light switch. Exactly how they do this is still unknown, but the significance is clear: each species has its own pattern of flashes, prefers specific habitats, and flashes only during certain periods of the night, all of which help members of each species to identify one another. There are many ways in which signals may vary: the number of flashes in a signal, the duration of the signal, the time interval between signals, the distance flown between signals, and even the color of the flashes.

While emitting synchronized flashes, male fireflies fly over an area likely to contain females. When the wingless, larvalike female (also known as a glowworm) recognizes the correct sequence of her species, she signals back. As they continue to signal each other, the male quickly closes in, lands, and copulates with her.

After observing flash patterns for a time, you may be able to duplicate them well enough to elicit responses from the insects themselves. Using a penlight with a sensitive switch, you can repeat the male's pattern downward from about chest height or the female's response upward from the ground. You might even capture one for identification and record the species and its flash sequence in your field notebook.

Above Larva of the *Lamprigera* sp. glowworm from the rainforests of Sumatra. Glowworms are predatory insects which feed on other invertebrates. The females glow to attract males.

SOCIAL INSECTS

By and large, most insects are solitary, indifferent to the company of others of their species except during mating and chance encounters. There are quite a few exceptions, however, and these have been lumped into one group called social insects. Termites, ants, most bees, and some wasps are true social insects. Many others will congregate for one reason or another at some point in their lives, but they do not stay together in organized societies throughout their life cycle.

Social insects live in cooperative, interdependent colonies of a single species, usually with a sharp division of labor among social *castes*. Such insects regularly care for the eggs, larvae, and pupae in the colony, a characteristic rare in the rest of the Class Insecta. They build nests of various degrees of complexity, and usually they have all descended from one female, the queen. They also practice *trophallaxis*, the mutual exchange of food and other desirable substances between members of the colony.

The evolution of social life among insects can be divided into several stages. Most insects do not live long enough to see their offspring, making parental care difficult. Probably the first significant step toward a social life occurred when solitary wasps and bees began to build nests, which they provisioned with food and in which they laid eggs. This practice, which still occurs among many solitary wasps and bees, represents a basic parent-offspring relationship, even though the two generations never see each other.

As their lifespans grew longer and began to overlap those of their offspring, the next step, also evident among some modern species, became possible. The mother lays her eggs in a nest with few or no provisions, but returns with prey periodically and feeds the larvae directly. Trophallaxis probably co-evolved as an inducement to this behavior, since the larvae of many social insects, when fed, are stimulated to secrete substances relished by the adults, providing the latter with a powerful incentive to provide food.

The beginning of a social structure is illustrated by certain solitary bees, which we call *sociable* because of their tendency to build their single- or multi-celled nests adjoining those of others of the same species. From this point, it is an easy

Above Worker termites (*Termites longipeditermes*) from Penang, Malaysia, carry in their mandibles balls of powdered lichen, collected from branches for their galleries. Collected vegetable matter is injected with a fungus, which the termites harvest when it has grown.

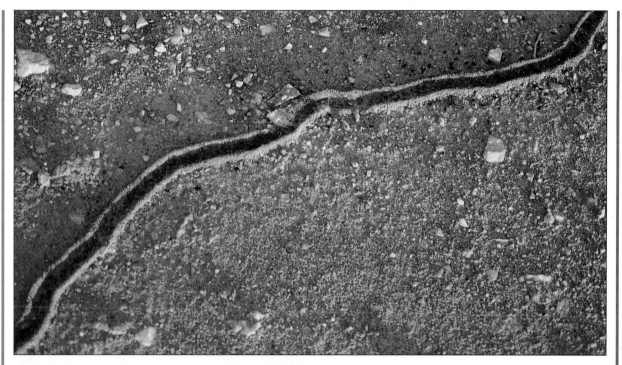

Above Creatures of habit with a strong group instinct, these driver ants (*Dorylus nigricans*) from Kenya have carved an ant highway with their continued traffic. Like other social insects the ants communicate in part with volatile pheromones, which elicit rapid responses when dangers threaten.

Left Worker ants of *Myrimica ruginodis* tend their larvae in a formicarium. One of the workers' main functions is to care for the eggs, larvae and pupae of the ant colony.

transition to a division of labor. Individual bees may adopt specialized tasks, such as guarding the entrance to the nest. While such action obviously benefits the insect's own offspring, those of the other bees also profit.

Caring for the offspring of another seems to violate the theory of the selfish gene discussed under "Insect Behavior," which asserts that the ultimate goal of (genetically programmed) behavior must be the survival of one's own genes. However, since all members of the colony descended from one queen, her offspring are basically siblings to all others in that society. Caring for them ensures the survival of any individual's genes, even if that individual did not contribute those genes directly. Of course, the bees themselves are not conscious of all this. They simply respond as their genes direct them.

Social castes

There are really only two major castes among social insects: the reproductive caste, composed of males and queens, which has the sole responsibility of making more offspring, and the non-reproductive caste, consisting of only workers or workers and soldiers, which undertakes all of the work in maintaining the colony, including building

and repairing the nest, collecting food, and caring for the eggs, larvae, and pupae.

Social bees and wasps have evolved an effective means of controlling the numbers in each caste to ensure that the colony operates efficiently. Only a limited number of males are needed to promote competition for the privilege of mating with the single queen. Extra queens leave with some males to start their own colonies. The queen has control over whether a given egg gets fertilized; unfertilized eggs develop into males, and fertilized ones become females. The queen, in turn, is stimulated to either fertilize eggs from sperm stored in her abdomen or withhold sperm based on the sizes of the cells built by workers. Into smaller cells she deposits fertilized eggs, while larger cells receive unfertilized eggs. Nutrition determines if a female grows into a worker, all of which are sterile females, or a queen. Among wasps, underfed or improperly fed females become workers, while well-nourished females grow into queens. Honeybee workers secrete a white paste, called royal jelly, from glands connected to their mouthparts and feed this to all larvae for at least their first three days. Those females that receive royal jelly throughout their larval stage develop into queens, and the rest become workers.

Right Social wasps (*Polybia occidentalis cinctus*) from Trinidad build a nest below a protective leaf in the rainforest. The nest is made entirely from wood fiber collected from tree trunks and chewed to make combs. The tiers of a wasp comb lie horizontally, compared to those of bees, which are vertical, and wasps remain on the surface of the comb to attack predators. The nest also has an outer protective covering.

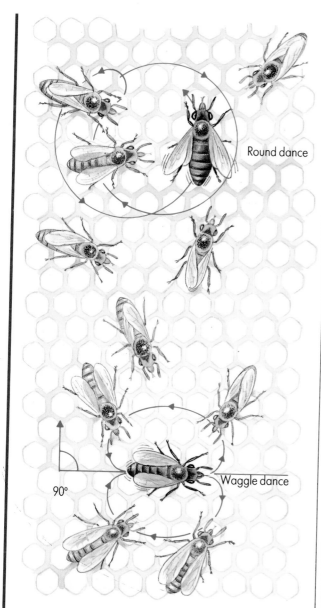

Round dance

90°

Waggle dance

Above Worker honeybees communicate information about pollen and nectar sources to members of their hive by "dancing." The locations of food sources within 80ft (25m) of the hive are communicated by a round dance. More distant food sources are indicated by a waggle or figure-of-eight dance, which involves waggling the abdomen from side to side during the straight portion of the figure-of-eight. Distance is indicated by the length of the run and the number of waggles. The angle of the straight run corresponds to the angle between the direction of the food source and the sun as viewed from the hive. The waggles and buzzes may indicate food quality.

What you can do

Make friends with a beekeeper

One of the best ways to learn the intricacies of social insects is to take up beekeeping as a hobby. Unfortunately, few of us have the time or finances to devote to this pastime, so the next best thing is to make the acquaintance of a beekeeper. Ask around at the farms, orchards, nature centers, or health food stores in your area and you will probably find out the names of several potential contacts.

Don't be bashful; most people love to share their hobbies or occupations with interested individuals, especially if they are good at it. They may even have extra protective equipment and invite you on their rounds. Do a little homework beforehand so that you are familiar with the terms they use.

Above Hands-on experience with honeybees, in the presence of a beekeeper, allows children to see brood, honey and pollen stores.

OFFENSE AND DEFENSE

The variety of the insect world has given rise to a great number of protection mechanisms. First and foremost is the insect's durable external skeleton, a veritable suit of armor for many species. Not only does it offer protection against physical forces, but it also prevents undesirable water loss and is resistant to a great many chemicals. In addition, the physics of having muscles *inside* the skeleton gives insects tremendous strength in relation to their small size.

Size itself is a great defensive asset of insects. Being small enables them to go about their lives unnoticed by many larger animals that would eat them. In fact, their diminutive stature is a major reason why entomologists estimate that there may actually be twice as many insect species or more on earth as those already classified – we may have simply overlooked the rest so far. Many just take refuge in a crack or under leaves or stones until danger has passed, and quite a few spend their entire lives out of sight of most vertebrates.

Of those that avoid danger by hiding, a great many rely upon some form of camouflage to enable them to hide out in the open. Some incorporate colors and patterns on their exoskeleton or wings to match the background material upon which they normally rest. Others have evolved fantastically modified body shapes and appendages that perfectly mimic plant parts such as twigs or leaves, right down to the swaying motion of a leaf in the breeze. Notable among these are the stick insects (Order Phasmatodea), leaflike katydids (Order Orthoptera), and treehoppers and leafhoppers (Order Homoptera).

Many of those insects that cannot hide so easily have another option – flight. As a group, winged insects have complete mastery of the air and can execute maneuvers beyond the capability of any other creature or machine. Of course, some are better fliers than others, but even the worst are better off than they would be if they could not fly at all.

Self-advertisers and mimics

In contrast to those which remain inconspicuous, a great number of insects boldly advertise their presence with bright colors or contrasting patterns. Most of those so marked have an unpleasant experience to offer potential predators, such as a

Right A female emperor moth (*Saturnia pavonia*) waits patiently on a grass stem, just after emerging from her cocoon. Her silvery colors, made up entirely of scales, portray a pair of fearsome "false-eyes" to deter predators. If the moth is disturbed, she will show off a further pair on her hind wings. She also gives off a powerful pheromone while waiting for males to gather round and compete for the chance to mate with her.

Left It is often the case that the smaller the moth, the more intricate the pattern. The coloration of this small moth, *Alabonia geoffrella*, has evolved as camouflage against predatory species.

venomous sting or spines, irritating hairs, repellent glands, or toxic or distasteful compounds in their tissues. Once so educated, an animal is unlikely to make the same mistake twice, and usually gives that species a wide berth. One might think that traits inviting attack would be quickly eliminated from the population, but apparently many more such insects are spared from repeat attacks than are lost to ignorant predators.

Riding on the coattails of these conspicuous insects are mimics, whose chance resemblance to a species habitually avoided by predators has encouraged their evolution to the point where it requires very close scrutiny to tell the harmless species from the dangerous one. Classic examples of this are the monarch and viceroy butterflies of North America. As a larva, the monarch, whose hazardous migration of thousands of miles each autumn has endeared it to millions of Americans, feeds exclusively on milkweed plants. Milkweeds produce a milky latex containing cardiac glycosides, toxic substances that cause nausea and vomiting in small doses and death to vertebrates in large doses. These toxins deter nearly all animals from grazing upon milkweeds, but monarchs have adapted to not only feed on them but to incorporate the cardiac glycosides throughout their tissues, making them extremely distasteful to predators, especially to birds that might otherwise pick them out of the air. Orange-and-black viceroy butterflies are a product of the same natural selection processes that govern monarchs; those that, through mutations, more closely resemble monarchs will more successfully pass on those traits to future generations.

Defense by offense

Social insects have adopted the axiom that the best defense is a good offense. While many insects are capable of delivering a painful bite or a venomous sting, the true power of this defense is unleashed when it is multiplied by hundreds or thousands of members in a colony. Social insects communicate extensively via pheromones, chemical messengers that elicit immediate instinctive responses. Alarm pheromones are potent and travel rapidly through a colony, rallying its occupants to repel any intruders.

What you can do

Gross grasshoppers

Equipment:
sweep net
bug box
hand lens

Many insects, when captured, give off a repulsive fluid to persuade their enemies to release them. The "tobacco juice" that a grasshopper spits as it is handled is really the foul-tasting contents of its gut. This is often effective in securing the insect's release, giving it just enough time to escape.

In summer, sweep an area of tall grass with your net until you catch a grasshopper. Holding it securely by both femurs will immobilize it for observation. Gentle prodding may encourage it to release a brown fluid from its mouth if it hasn't already. What does it smell like? Handle with care, as this substance will stain skin and clothing.

39

ENEMIES AND ALLIES

Insects have not had a fair deal. They are not popular subjects of nature study, in part because we tend to associate them with nuisance and with the transmission of diseases. As the most numerous and diverse class of organisms on earth, they are inextricably woven into the web of life, and so there are inevitably some adversarial relationships with humans. However, we also have a good many insect allies.

Beneficial insects

For one thing, insects are responsible for perpetuating many of the green plants that feed our planet and oxygenate its atmosphere. With the evolution of flying insects, plants were quick to take advantage of these reliable pollinators. Insect-pollinated plants have incorporated colors, fragrances, and nectar attractive to various insect species into their flowers, and many have even adapted to open their flowers only during the times of day when their primary pollinators are most active. Bees, butterflies, moths, and flies are the major pollinators. Hives of honeybees, perhaps the world's premier insect pollinators, are therefore frequently kept in orchards or nurseries, and on farms. As a side benefit to their pollinating activities, the bees supply us with large quantities of honey and beeswax, which is used in lubricants, salves, ointments, furniture polish, candles, and other products.

Other insects also benefit plants and, indirectly, us. The activity of burrowing insects, such as ants, loosens, aerates, and fertilizes the soil, encouraging the development of healthy plant roots. Some species feed upon noxious weeds, helping to control them, and predatory insects will readily take any insect within their power, including those damaging to plants. The value of insect predators, such as ladybird beetles, mantids, and lacewings, is legendary.

Insects have supplied us with a good many other products besides honey and beeswax. Most of the world's silk, for instance, is produced from the cocoons of the silkworm moth, *Bombyx mori*. Sometimes the insects themselves become the product; among certain cultures, insects, such as grasshoppers, flying ants and honey ants, are popular fare.

Harmful insects

Despite the fact that most insects have no direct interaction with humans at all, there are some that are unquestionably harmful. Mosquitoes can transmit such diseases as malaria, yellow fever, encephalitis, and elephantiasis. Fleas are the vectors of bubonic plague, typhus, and tapeworms. Houseflies have been implicated in the spread of tuberculosis, typhoid, cholera, amoebic dysentery, anthrax, and other illnesses. Human lice, though they are basically parasites that usually cause only discomfort, also may transmit typhus and relapsing fever. One of the best known disease vectors is the tsetse fly, which transmits African sleeping sickness.

Ever since the dawn of agriculture, people have done battle with insects over the crops that both

Above Silverfish are viewed as pests because they damage books and papers. This member of the silverfish family (*Lepisma saccharina*) reveals the silvery color responsible for its name.

Above The gardener's friend – ladybird beetles feeding on aphids.

Left A small swarm of "killer" bees hangs patiently in the branches waiting for "scout workers" to find suitable accommodations. "Killer" bees are a hybrid, the result of breeding between stocks of African bees and Latin American bees (*Apis mellifera*). Their aggressive nature is notorious.

desire; so far, the conflict seems to be a draw. The use of chemical pesticides during the 20th century has given humans the edge, but their use must be carefully regulated to avoid worldwide environmental disaster.

Insects may destroy crops in one of two ways. Most obviously, their consumption of the plant means there is less for us. Grasshoppers, true bugs, aphids, and beetles are the most serious offenders in this category. Infestation of our stored grains and cereals is also a serious problem, but most of this damage is caused by only a few species of beetle. Perhaps more serious are the many plant diseases introduced through the wounds where insects have been feeding, boring, or laying eggs.

Adding insult to injury, certain insects inflict considerable damage on our possessions as well.

Termites are a worldwide threat to any wooden structure. Carpenter ants and wood-boring beetles pose a similar but much lesser problem. Silverfish (Thysanura) and book lice (Psocoptera) can destroy books and papers as they feed on paste, glue, and the sizing used to create glossy pages. Clothes moths were once significant destructive pests of natural fabrics, but their significance diminished with the increased use of synthetic fibers. Carpet beetles (Family Dermestidae) are a much more serious problem today, as they attack both natural and synthetic fibers.

Despite the serious nature of insect pests, it behooves the practical entomologist to dispense with any prejudice that may interfere with the study of this fascinating field of natural history. Bear in mind the many beneficial insects, and the fact that most insect species are basically benign.

ENTOMOLOGY IN ACTION

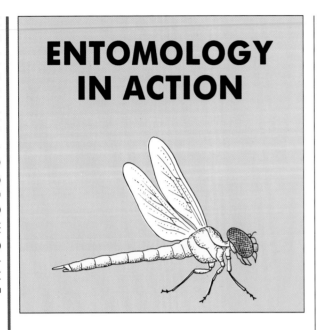

TOOLS OF ENTOMOLOGY

The beauty of entomology as a pastime is that you can find insects almost anywhere and at any time. You can spend a few minutes here and there or a whole day, depending upon your level of interest and time available, and with such a wide array of subjects, you need never be bored. Best of all, you need very little specialized equipment to get started. Even so, a few appropriate tools will yield much more for your efforts.

Tools for examining specimens

Since adult insects rarely stand still, you will need to confine live specimens in some way in order to study their features. Any wide-mouthed clear glass or plastic jar will do, but the smaller ones are generally more suitable. Especially useful are bug boxes, which are clear plastic cubes with a magnifying lens in the lid. These come in a variety of sizes and can often be purchased in the gift shops of natural history museums and nature centers. If your subject is particularly active, you can slow it down by placing it in the freezer for a few minutes.

A good quality glass hand lens, available from biological supply companies, will reveal minute details of insect anatomy that you may otherwise overlook. A magnification of 8x to 10x is optimal, and some models have several glass elements that can be used in different combinations to vary their magnifying power. Photographic loupes sold in photography shops will also serve well.

Despite the disclaimer that practical entomologists do not need expensive equipment, the author does recommend one such instrument for those with a serious interest in natural history. A stereomicroscope, also called a dissecting or binocular microscope, is a wonderful tool that provides an unequaled three-dimensional view of small objects. It will usually have a variable magnification from 10x to 60x, and is invaluable for viewing or dissecting insects. Several models are available through biological supply companies, and while they are not cheap, most are affordable, and nearly all cost less than good-quality conventional light microscopes.

Examining insects against a contrasting background will often help you to see their finer features. Keep small pieces of black and white cardboard, about four inches square, with your equipment for this purpose. A small penlight will be useful under low light conditions.

Tools for handling insects

There are several implements that are handy for manipulating captured insects. Forceps – similar to tweezers but with a more sensitive touch that is less likely to damage your specimens – are useful in handling specimens. Probes may be used to prod an insect into moving in a certain direction or to spread the wings and legs of killed insects for examination. Both probes and forceps are easy to make and are also available from biological supply companies. To move small killed insects, a camel hair artist's brush is ideal.

Tools for dissecting insects

You will find a dissecting tray handy for examining or dissecting killed insects. You can easily make your own by pouring melted paraffin into a small baking tray to form a layer of wax about one centimeter deep. When the paraffin becomes too full of holes or gouges, simply melt it again and let it solidify. Pins of assorted sizes are useful for securing insects during dissection, and can be ordered from biological supply companies.

Should you wish to dissect any specimens, you will need cutting instruments. Scalpels are available commercially, but expensive. A good substitute is an X-acto razor knife with interchangeable

blades, available in hobby stores. An old travel toothbrush holder makes a safe carrier for it. A fine pair of scissors may also prove useful.

Field guides

Invaluable in the study of most other fields of natural history, field guides are of more limited worth in entomology, primarily because no single guide can list even half of all insect species in a given region. Even so, the insect field guides that are available usually list at least the most common species in the geographical areas they cover, and so deserve a place in your library.

Above There are three main reasons for capturing insects: to aid in identification, to study their behavior and life cycles, and to create a collection. There are a number of methods you can use, but in all cases the tools required are comparatively simple, ranging from nets to beating trays and traps.

All good field guides have several features in common. They should be logically organized, preferably by order and family rather than by some nebulous criterion, such as color, size, or habitat. Detailed illustrations are a must, as are range maps or descriptions, specific habitat preferences, and clear physical descriptions of each insect, including characteristic identifying features. Some also list interesting facts about each individual species, such as its life cycle, behavior, and significance to humans.

Identification keys

Useful in conjunction with field guides are identification keys, sometimes called dichotomous keys, that guide you with certainty to the order and family of your specimen. At each step, the key will present two options, only one of which will apply to the insect in question. Whichever one you select will either reveal the order or family of your specimen or guide you to the next step in the key. Keys to insect orders and families are not as common as field guides, but you may find them in certain field guides, in trade books sold in book stores, or in entomology textbooks at your local library. An entomology professor in any university may also be willing to make copies of keys for you.

Keys to individual insect species would be cumbersome, and the features separating one species from another are often so minute that they are impractical to discern. Fortunately, the nature of entomology is such that knowing the order and family of an insect will provide plenty of clues about it.

Recording what you see

In many respects, the most useful tool of any practical entomologist is a well-kept field notebook. Knowing the name of a species is no prerequisite for making observations of its appearance and behavior. You can even make up your own name for a specimen as you take notes, thus giving yourself a handle on that species until you can make a positive identification.

Look in stationery stores to find a medium-sized notebook, strongly bound and composed of high-quality, unlined paper that will not tear easily. Keep it as clean as possible to preserve its usefulness. Self-sealing food storage bags make tough, waterproof storage cases for your notebook, field guides, keys, and maps while in the field.

Keep a pen and several pencils in your day pack along with the rest of these tools. Backpacking and outdoor shops sell space-age pens that write in any position and under any conditions. A short ruler is also handy for recording dimensions.

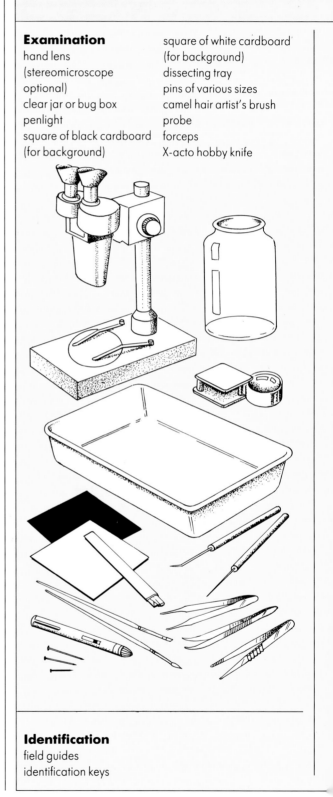

Examination
hand lens
(stereomicroscope optional)
clear jar or bug box
penlight
square of black cardboard
(for background)
square of white cardboard (for background)
dissecting tray
pins of various sizes
camel hair artist's brush
probe
forceps
X-acto hobby knife

Identification
field guides
identification keys

Tools of the practical entomologist

Recording
field notebook
pens and pencils
ruler
camera and accessories

Collecting
aerial net
sweep net
dip net or kitchen strainer
small white plastic or metal
tray
aspirator
killing jar

specimen box
relaxing chamber
triangular paper
envelopes
insect pins of assorted
sizes
pinning block
spreading board

display boxes
paper points and labels

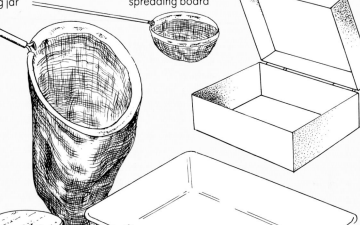

Miscellaneous
day pack
topographic maps of local
area
compass
Swiss army knife
water bottle

45

CAPTURING INSECTS

There are many methods of capturing insects, and any one is acceptable as long as it is safe and does not damage the insect. Many insects are easily collected by hand, with forceps, or by simply clapping a jar over them. Others require a bit more finesse and some basic tools. The following are a few such devices that are either simple to make or else inexpensive to buy.

Nets

Many insects are adept at evading or escaping predators and can easily evade our clumsy efforts. Others can be dangerous to handle; bees and wasps inject venom as they sting, and many of the larger true bugs can deliver a painful bite. For these, and for insects that hide in tall grass or underwater, we need a net, of which there are three basic types.

An aerial net, with which most of us are familiar, is used for netting flying insects out of the air. It is

Above The choice of a suitable habitat for an invertebrate or an insect depends on knowing the ecological requirements of the species, and spending time in the field helps one to understand various groups. It is essential to be prepared and to take along the right equipment.

a long sack of soft nylon, organdy, or silk netting that will not normally damage specimens. This delicate material does not wear well and should not be used to capture insects in vegetation or on the ground. After netting an insect with an aerial net, give the handle a quick half-twist to double the net over the opening and prevent escape. The best way to retrieve your catches is by inserting a jar into the net and trapping the insect between the netting and jar.

Sweep nets are made of heavy white muslin or canvas and are used with a sweeping motion to capture unseen insects in tall grass or weeds. This "shotgun technique" rarely fails to turn up something of interest. It is especially useful in meadows during the summer and fall.

A dip net is a long-handled tool used to capture aquatic insects. It has a flat side that may be placed against the bottom of a stream or pond. The triangular hoop is threaded through a tube of heavy muslin with a screen of heavy nylon netting on the end, to allow water to flow through. Insects may be difficult to see among aquatic debris, so you may need to dump the contents into a white tray and watch for movement.

Beating tray

Insects that frequent plants are often reluctant to fly or run, preferring instead to remain still and rely on their natural camouflage and small size to escape detection. A sweep net, good for grassy meadows, does not work as well on trees and shrubs; here you need a beating tray. This is a simple affair made of white cloth stretched over a frame of two crossed sticks. (A light-colored umbrella also works well.) It is held or placed under a shrub or tree branch, to which you deliver several sharp blows, taking care not to damage the plant. Insects having taken refuge there will fall onto the beating tray and can then be collected by hand or other means.

Pitfall traps

Pitfall traps are effective in catching insects that scurry across the ground rather than fly. Simply dig a small hole with a garden trowel and sink a jar or cup in it so that the top is flush with the surface of the ground, filling in with soil around the outside of the container. An inch of 30 per cent water to 70 per cent alcohol mixture in the bottom will kill the insects if you do not wish any to escape. If you do

not want to kill any insects, use yoghurt pots with drainage holes and check every few hours. Place three or more strips of wood radiating from the rim like the spokes of a wagon wheel; these will guide insects into the trap. Cover the trap with a slab of wood or stone to keep out rain water, and check it daily. You should set several traps in a variety of locations in order to find the best sites in a given area.

Bait traps

One of the major pastimes of insects is searching for food, and many species can be lured with bait. Pitfall traps or small containers suspended from a tree branch make excellent bait traps. Cover them securely with a large-mesh screen with openings about 2 cm square to admit large insects but keep out other scavengers.

A mixture of equal parts of molasses and water makes a fine bait. Add a little yeast and allow the

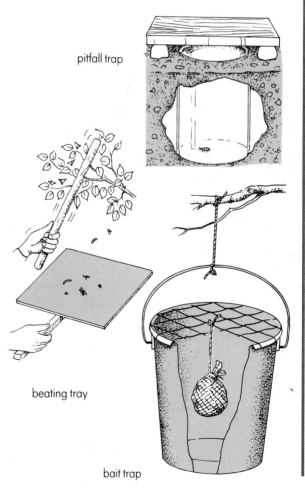

pitfall trap

beating tray

bait trap

mixture to ferment for a day or two before baiting traps with it. A mash of fermenting fruit also works well. Cloth strips soaked in a bait mixture and suspended from a tree branch will attract larger insects, such as nocturnal moths, that cannot fit through the screen covering your other bait traps. The mixture can also be painted onto the bark of a tree. Visit the site every hour or so during the night.

Decaying animal remains or meat scraps will attract many species, especially beetles and flies. If you wish to kill your prisoners, suspend the bait in a cheesecloth sack from the wire screen and put an inch of the water and alcohol mixture in the bottom of the trap. Dung is also a very effective bait, but take precautions against the transmission of disease or parasites by handling it only with disposable implements. The water and alcohol mixture in the trap should sterilize any specimens attracted to the dung.

Berlese funnel

More insects than you might have guessed reside in the cool, moist debris littering a forest floor. A Berlese funnel is a device that can drive insects from such material. Simply place a large funnel, one that narrows to about a two centimeter opening, small-end down in a tall glass jar containing a layer of moistened paper towels. Fill the funnel with leaf litter and suspend a high-wattage light bulb over it. As the debris warms and dries, any insects it harbors will burrow deeper,

seeking their preferred conditions, until they reach the funnel opening and fall into the jar.

Sifter

A sifter will do much the same job as a Berlese funnel. To make one, cut the bottom from an old bucket, leaving about a one-inch lip. Cut a circle of large wire screen to rest in the bottom of the bucket. Fill the sifter half-way with debris and cover with canvas, securing the cover with cord or a large rubber band. Shake the sifter vigorously over another bucket or other deep container and collect your specimens.

Light trap

It's common knowledge that nocturnal insects are attracted to light, and you can use this behavior to your advantage. There are many different types of light traps. Perhaps the simplest is a white sheet tied at all four corners and suspended vertically, with a lantern shining on it from behind. As insects alight on the sheet, they may be simply scooped off with an aerial net, allowing you to capture only the ones you want. An empty Berlese funnel with a cylinder of wire screen extending up around the light bulb also makes an effective light trap, though you may need to adapt it by using a funnel with a larger opening at the bottom.

An aquatic light trap can be made with a flashlight, a watertight jar, a length of pipe about six inches in diameter, and some fine-mesh wire

berlese funnel

light trap

Top Fogging is a technique now widely employed in rain forest in order to find out which insects live in the canopy. A large gun is hoisted aloft, armed with burning insecticide and a petrochemical, and maneuvered around at first light, disturbing the sleeping insects which then fall to the ground.

Above The rain forest floor is festooned with large canvas funnels for collecting insects. Fogging is carried out when there is little wind so that everything falls into the funnels instead of being driven elsewhere.

screen. Turn on the light; seal it in the jar, and place it in the pipe. Use the wire screen to cover the end of the pipe toward which the light does not point, and fashion a funnel out of the same material for the other end, pointing inward. Secure the pipe with strong cord and lower it into the water, with the funnel facing upstream if there is a current. You may need to weight it down with rocks if the pipe is not heavy.

Aspirator

Some insects that you will find are so small that they cannot be easily collected by hand or with a net. For these, the best method may be the use of an aspirator, with which you manually suck the insects into a small vial.

You can make an aspirator from any small vial with a tight-fitting cap or stopper, two plastic straws, and a length of flexible plastic or rubber tubing. Drill two holes, exactly the same diameter as the straws, in the cap. Insert one straw so that it is about one centimeter above the bottom of the vial. Cut a short piece from the second straw and insert it so that about two centimeters extends into the vial and one centimeter protrudes above the cap. Seal the joints between the cap and straws with silicone bathtub caulk. Cover the inside of the short straw with a small piece of fine netting so that you don't accidentally suck insects into your mouth. Attach one end of the rubber tubing to the other end of the short straw, seal with silicone caulk, and you're ready for action. Turn over stones or sift through forest and plant debris to suck up little insects like springtails.

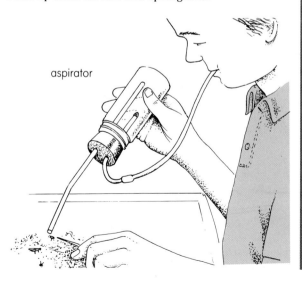

aspirator

KEEPING LIVE INSECTS

Observing the behavior and life cycles of insects provides fascinating insights into their lives, and one of the easiest ways to do this is by watching captive insects. A simple jar with a ventilated lid will serve as a cage for some species, while others may require more upscale accommodation.

If you wish to keep your subjects alive indefinitely, you should identify the species you are dealing with in order to research their needs. At the very least, you will need to determine their orders and families to make an educated guess about their preferred conditions. Like all animals, insects have certain basic requirements: food, water, and protection from extremes in their environment. In order to raise more than one generation, you will also need to provide suitable sites in which they can pupate and lay eggs. Bear in mind that species which overwinter as eggs or pupae may fail to develop if not subjected to low temperatures. You can satisfy this requirement by placing the eggs or pupae of such species in the freezer or outdoors for a few weeks.

Probably the most specific need of your captive insects is food. Water is extracted by some insects, especially plant-eaters, from their food, and most other terrestrial insects thrive on a light misting from a spray bottle, which simulates dew. Be careful, though; large droplets can trap small insects, and too much moisture encourages the growth of mold. As for environmental conditions, the climate in any home is moderate enough for nearly any insect. However, many insects are specialists with precise food requirements. This is particularly true of the herbivores, which often have evolved life cycles centered around a specific plant species that is reliably found in a given habitat. A good example is the monarch butterfly, *Danaus plexippus*, whose larvae eat only milkweed plants. Fortunately, there are plenty of other insects which are easy to satisfy. Scavengers and predatory insects are usually less picky about their foods, since in nature they have evolved to take advantage of whatever comes along.

Terrestrial insect homes

The style of cages you choose for your insect zoo is limited only by your imagination, though they must be ventilated and have lids or doors that

Above Milkweed bugs (*Oncopeltus fasciatus*) mate in characteristic fashion on a leaf. These are from St. Vincent in the West Indies and can walk around in broad daylight in defiance of predators, since their bodies contain poisonous substances gathered from the plants that they feed on; this protection is advertised by their warning colors.

Left Mealworms (*Tenebrio molitor*) are the larvae of darkling beetles, and are often reared in laboratories for study of insects, and as bait for fish. The larvae are pests of cereal, grains, bran, and other stored foods, where they eventually pupate and emerge as winged adults.

close securely. It also helps if they are easy to clean, which should be done periodically. Attempt to simulate natural conditions if you wish your guests to remain alive for a substantial period. Include some soil, sand, or rocks if they are part of the insects' natural niche, and perhaps a stick or plant for them to rest on or climb.

Large, wide-mouthed jars with mesh covers or ventilated lids are simple yet effective cages. If space is not a problem, an aquarium with a snug-fitting screen top also works well. A wire colander from the kitchen makes a good home when turned upside down, provided there are no gaps around the bottom.

A versatile insect cage may be made from a piece of wire screen and two cake pans of equal size. With a tape measure, determine the circumference of one of the cake pans. Cut a rectangular section of wire screen with one dimension equal to this measurement and join the two sides with duct tape or monofilament fishing line to form a cylinder. Rest this inside one cake pan, and use the other for the lid. A pane of glass or plexiglass may substitute for a lid if the screen is sturdy enough to support it.

Insects that live in plant litter can be reared by placing the debris in a closed box with a vial protruding from one side. As adults emerge, they

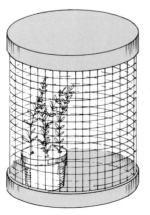

Insect cage made from wire screen and two cake pans

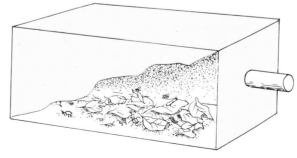

Insect cage made from a vial inserted in the side of a cardboard box

move toward light and will end up in the vial.

Unless you supply plants for food, maintain moderate humidity near terrestrial insects by keeping a piece of moist sponge in each cage. Plants provided for food must be kept fresh or replaced regularly, and potted plants are a better food source than cuttings. If plant cuttings are placed in water, cover the water container, leaving holes large enough for the stems, but small enough to prevent the insects from entering and drowning.

An alternative to simulating natural conditions in your home is to cage herbivorous insects in the field. If you find insects, immature or adult, feeding on a small plant, you can cover the whole plant with a screen cylinder and a lid and visit it periodically to observe them. If they are feeding on a large plant, tie a fine-mesh bag around that portion of the plant.

Aquatic insect homes

Aquatic insects, as one might expect, require an aquarium or a large glass jar. Fill the aquarium with water from the place where you capture the insects, not with tap water. Aerate the water with an air pump if the insects come from a stream; otherwise, this will probably not be necessary. Line the bottom with clean gravel or coarse sand for easy maintenance, and slope it to above water level to provide shallow and deep areas. Most adult aquatic insects can also fly, so a wire screen or glass lid will be needed. Some may need plants or sticks emerging from the water so that they can crawl up and make their final molt into adulthood.

You may wish to create a balanced aquarium by introducing some small native fish, amphibians, other invertebrates, and plants as well, but remember that a crowded aquarium will not stay healthy, and an excess of plants will make the insects harder to see.

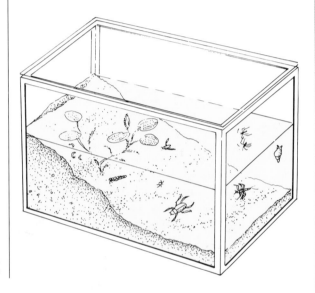

Above An insect aquarium will enable you to observe the life cycle and behavior of aquatic insects, such as dragonflies, caddisflies and mayflies. Clean gravel, plants, and water taken from the capture site will help to simulate suitable natural conditions.

Left This fresh water aquarium could easily be adapted to provide a home for local aquatic insects.

Above This four-spotted chaser (*Libellula quadrimaculata*), photographed in England, has just emerged from its molted exoskeleton, which still hangs on the vegetation. The chaser expands its wings until they are hard and firm, ready for flight. Like that of many newly-hatched insects, its body is clothed in a fine layer of hairs. Note the predator's characteristic large compound eyes with thousands of ommatidia, jaws, and legs suited to catching prey in flight.

53

MAKING A COLLECTION

Creating an insect collection is not necessary to practical entomology. You can learn a lot by observing the behavior of wild or captive insects, and by examining specimens that you have captured without harming them. Building a worthwhile insect collection is a meticulous and time-consuming endeavor. However, naturalists seem to have a penchant for collecting things, and so a portion of this chapter is devoted to helping you assemble a creditable insect collection.

If you decide to start an insect collection, it should be assembled in such a way as to be of value to the scientific community should you ever decide to donate it to a museum or university. Each collection should be composed of correctly-mounted and well-preserved specimens, and it should also tell a story about where, when, and by whom each specimen was collected. You may choose to specialize and collect only a certain order of insects, such as beetles.

A few words of caution: do not add butterflies or moths to your collection unless you are serious about maintaining it. The Order Lepidoptera is the one major group of insects whose members have shown any serious population declines in the face of human activities. If you do collect butterflies and moths, know the endangered species in your area and refrain from killing or injuring them.

Killing insects

Dispatching insects for your collection requires a killing jar with an airtight lid into which you place the insects, together with a volatile, poisonous substance. Professional entomologists sometimes use potassium cyanide as a killing agent, but this is extremely toxic and very dangerous if not handled with extreme care. By far the better choice is ethyl acetate, which is available from pharmacists and biological supply companies and is far less hazardous. If ethyl acetate is not readily available, acetone (fingernail polish remover) is a good substitute. No matter which of these two you choose, avoid breathing the fumes or spilling any on your skin. Killing jars should be clearly labeled "POISON" and should be kept out of the reach of children.

Killing jars are easy to make. You will need three wide-mouthed jars – large, medium, and small –

with screw-top lids. Pour about a one-half inch layer of plaster of Paris, made with a little extra water so it pours easily, into the bottom of each jar and allow them to dry overnight with the lids off. Tape the bottom third of the jar with electrical tape to reduce the chance of its shattering if dropped. Just prior to using each jar, sprinkle a few milliliters of ethyl acetate onto the plaster of Paris, which will absorb it, and close the lid for about five minutes. You will need to recharge the jar with ethyl acetate every few days, depending on how often it is opened. Reserve the large jar for

butterflies and moths, which tend to lose wing scales that could foul other specimens. Wipe the jars clean periodically.

Transporting specimens

Rarely will you mount specimens in the field, so you will need some means of transporting them without damage. The killing jar can serve this purpose if there are only a few insects. Otherwise, purchase an ordinary small plastic food storage container with a snap-on lid; line this with absorbent cotton, and sandwich your specimens

Above This cabinet with glass-topped drawers displays insects in a traditional manner, enabling comparisons to be made. Familiar to Victorian naturalists, the insect cabinet is now rare.

between thin layers of cotton. Place each butterfly and moth in an individual triangular paper envelope first, in order to protect their wings (see illustration on p45 in "collecting tools"). Insects will dry when stored in this fashion and must subsequently be softened in a relaxing chamber before mounting.

Relaxing chamber

Dry insects become brittle and nearly impossible to mount. Overnight storage in a relaxing chamber, which is an airtight container of high humidity, restores their flexibility, so that legs, wings, antennae, and other parts are less likely to break as the specimen is handled during mounting. Make a relaxing chamber from the same type of airtight container as is used to transport specimens in the field, lining it with moist paper towels or a damp sponge. Rest the insects on a small dish inside to keep them from direct contact with the moisture, and place a separate container inside the chamber with a few milliliters of ethyl acetate or a few grams of moth flakes to inhibit the growth of mold.

Mounting specimens

In order for your collection to be worth anything, the specimens must be mounted properly. In most cases, this entails pinning the insect through the thorax so that it is suspended right side up, and parallel to the floor of the collection case. Special insect pins, available through the biological supply companies listed in the appendix, must be used to mount your specimens. These are available in different sizes, No. 1 being the most slender and No. 5 the thickest that you will need. They are usually black and coated with varnish to make them rustproof. Common sewing pins will damage specimens because they are too thick, and they will rust.

Most insects, including flies, wasps, bees, butterflies, moths, mantids, and cockroaches, should be pinned vertically through the right side of the thorax between the bases of the front wings, with the pin emerging where it will not damage legs or other delicate identifying features on the underside. Beetles and most families of the Order Homoptera are pinned through the front portion of the right wing. True bugs should be pinned through the right side of the scutellum if it is large enough to do so, or through the right wing like beetles if it is not. Grasshoppers, crickets, treehoppers, and leafhoppers are pinned through the right rear portion of the pronotum; grasshoppers may also be pinned with one wing spread on a spreading board. Dragonflies and damselflies may either be pinned vertically, with their wings spread, or horizontally through the thorax, with the left side up and with their wings positioned over

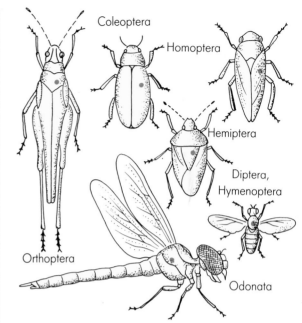

Above It is important to mount specimens correctly. The insect should be pinned through the thorax so that it is suspended right side up, and parallel to the bottom of the case or drawer. The position of the pin will vary depending on the type of insect.

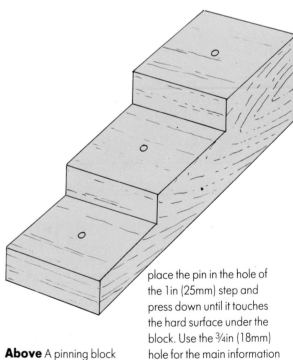

Above A pinning block helps you to pin specimens at a uniform height. Once the pin is through the insect, place the pin in the hole of the 1in (25mm) step and press down until it touches the hard surface under the block. Use the ¾in (18mm) hole for the main information label, and the ½in (12mm) hole for any secondary label.

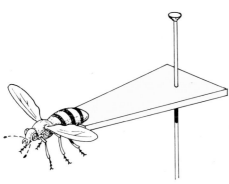

the back before pinning. Long abdomens, antennae, and other structures that might droop can be supported by pushing the pin deep into a soft surface, such as cork, balsa wood, or styrofoam, positioning the parts as they rest on this surface and leaving them until they dry.

Pinning block

Mount specimens by holding them between the thumb and forefinger of one hand and inserting the pin with the other. All specimens should be pinned at a uniform height to a pinning block, available from biological supply houses. The most useful pinning block has steps 25mm, 18mm and 12mm high, with a small hole drilled vertically through the middle of each step. Once the pin is through the insect, place the pin point in the hole of the 1in step and push it down until it touches the hard surface upon which the block rests. With thick-bodied insects you may need to withdraw the pin somewhat to leave space enough at the top for easy handling.

Mounting small insects

Some insects are so small that pins would damage their delicate structures. Wide-bodied insects less than 6mm long or narrow-bodied specimens less than 9mm in length are best mounted on points, small triangular pieces of thin white cardboard made entirely from rags so that they will not deteriorate with age. Special punches to make points may be purchased from biological supply houses, as can the points themselves, or you can make your own by cutting out isosceles triangles 4mm at the base and 8mm high. Bend the tip of the point slightly downward with forceps and use a very small amount of white glue or clear fingernail polish to attach the right underside of the insect to the tip of the point, trying not to obscure many of its features. The point can then be positioned on a pin, using the 1in step of the pinning block.

Spreading board

Butterflies and moths must be pinned with their wings fully spread, as these usually display the most important identifying characteristics. Dragonflies and damselflies are mounted in a similar fashion, and grasshoppers may have one of their wings spread to show their detail. A spreading board has a groove or slot in the center to accommodate the body of the insect, and two

Right Small insects will be damaged by pinning. Instead glue the right underside of the insect to a point (cardboard made entirely from rags) and then position the point on a pin.

Below Butterflies, moths, dragonflies and damselflies should be mounted with their wings spread. First pin the insect and then place the pin in the cork strip at the bottom of the central groove. Position the insect's body in

the groove, and fasten the wings to the spreading block with pins and strips of paper until the insect is completely dry.

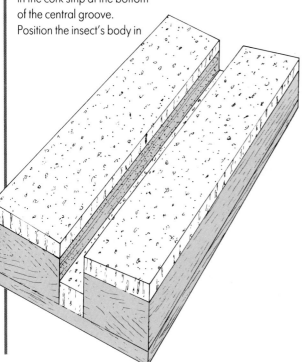

slightly sloping sides on which the wings are positioned with pins and strips of paper as they dry. This gives the wings a slight upward tilt, allowing them to settle approximately to a flat plane as they age. Specimens should be pinned before their wings are spread, and you will find a cork strip at the bottom of the central groove to accept the pin on which the insect is mounted. Wings of butterflies and moths should be arranged so that the rear edge of the forewing is perpendicular to the body. Drying times depend upon the humidity, but generally vary from two weeks for smaller insects to one month for larger ones.

Alcohol preservation

Some insects are simply too small to mount even on points, and others have soft bodies that would wilt if mounted by conventional means. These are best preserved in small, tightly-capped vials containing a 70 per cent solution of alcohol. Several members of the same family, collected at the same time and place, may share a vial. Replace the alcohol once about a week after introducing the insects. Vials with leak-proof caps can be secured horizontally in display cases with the rest of your collection, but they may need to be recharged with alcohol occasionally, as small amounts may evaporate over time.

Labeling specimens

Unlabeled specimens have very little value to the scientific community. At the very least, each specimen should be labeled with the location and date of its capture and the collector's name. Location details should include the country (abbreviated), state (abbreviated), and county or nearest town. Record the date with the month abbreviated (for example, Mar. 4, 1991) rather than written numerically (3/4/91), which could be mistaken for April 3, 1991. If your full name is too long to fit on the label, it may be recorded by your first two initials plus last name. Use the 18mm step on your pinning block to position this label parallel to the insect's body (head-to-toe axis) with the pin in the center of the label.

Additional information that would add to the value of your specimen may be recorded on a second label. Such data may include the collector's name (if this is not on the first label), and the insect's habitat, the method of capture or equipment used, the elevation at which it was found, or

Above A particularly fine example of a glass-fronted insect cabinet; now a collector's piece, this is the best possible way of preserving a collection of insects. The moths seen here are displayed in species rows, each specimen carrying a label of name, date, and locality. Every so often the drawers must be filled with an insecticide to deter the museum beetle, which can decimate collections in a few months.

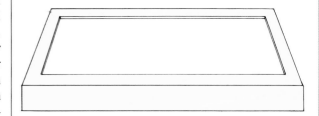

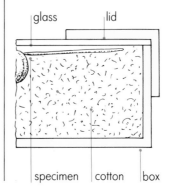

glass   lid

specimen   cotton   box

Above and **left** Pinned insects should be kept in a box or case with an airtight lid. There are a number of models available including this Riker mount. Here, pinned specimens with spread wings are placed upside down on cotton just below the glass.

Right The French naturalist, Jean-Henri Fabré – frequently called "the insect man" – collected plants and animals, but especially insects. He was a contemporary of Charles Darwin and lived in Provence, France. Today his collection can be seen just as he left it. His study has numerous boxes filled with local insects from the South of France and from the Massif Central.

a specific landmark. This label should be placed parallel to the first label and underneath it, using the 12mm step on your pinning block.

Each of these labels should measure 8.5mm × 17mm This is a fairly small area, so you must print very small, preferably with an extra-fine point pen and permanent ink, such as India ink. Use stiff white paper, with 100% rag content, that will not deteriorate.

If you are certain of the specimen's identity, you may record the species name, the name of the person who identified it (preceded by "det." for "determination"), and the year in which it was positively identified. This data should appear on a label measuring 25mm square and located on the pin on the floor of the display case.

Caring for your collection

A collection in which you've invested valuable time deserves a good home. This should be a flat case, preferably wooden, with a snug lid that is as airtight as possible and closes securely. Various models are available through biological supply companies, and as your collection grows, you might want to consider buying a cabinet into which display cases fit as drawers. If you are fortunate enough to know a talented woodworker, ask that person to construct a display case for you, preferably with a large glass window in the lid. Weatherstripping can be used to keep in fumigants

and keep out dust and insect pests that feed on specimens. Line the bottom with soft, white cardboard over cork or polystyrene to accept pins.

Collections should be arranged by orders, and by families within each order. Orders should be arranged from the primitive to the most modern, as determined from your field guides. You can purchase or make individual unit trays from white cardboard to accommodate separate orders, or even families in larger collections. A large label with the Latin and common names of each order may be pinned to the floor of the case, with a label listing the Latin and common names of each family pinned below the order label. You may wish to color code the ink or paper on order and family labels for display purposes. Arrange specimens of each family in orderly rows and columns.

Pests such as dermestid beetles that feed upon dry specimens can be repelled by including fumigants, such as naphthalene or para-dichlorobenzene (PDB), both sold commercially as either moth balls or moth flakes. A combination of the two works well, as PDB is more toxic to pests but naphthalene lasts longer. Place them in small boxes with screened lids in the corner of your display case, and check them every month or two. If an infestation does occur, as evidenced by sawdustlike material beneath specimens, or shed beetle larva cases, introduce ethyl acetate on a wad of cotton or gauze to the closed display case.

INSECT PHOTOGRAPHY

A challenging and rewarding alternative to keeping an insect collection is insect photography. For practical entomologists, there is only one group of cameras to seriously consider; the 35mm single-lens reflex (SLR) models *with interchangeable lenses* offer by far the best compromise between quality, versatility, convenience, ease of operation, and cost. These are the first choice of nature photographers around the world.

The quality of photographs taken by a competent photographer using a 35mm SLR camera is difficult to beat, and its versatility is second to none. With it, possible subjects range from the microscopic to the astronomical. By simply changing lenses, you can go from shooting close-ups of a beetle to taking telephoto shots of a skittish songbird in a matter of seconds. They are also light enough to go anywhere. Their ease of operation varies, but to use the newest auto-focus, auto-exposure cameras is literally as simple as point-and-shoot. Finally, the cost of 35mm cameras and basic accessories is well within the budget of most people. Second-hand cameras in good condition can often be found in camera shops or through classified advertisements, but it is wise to have any second-hand equipment examined by a qualified camera technician before purchasing it.

Choosing a brand of equipment is largely a matter of personal preference. For background information, write to the major photography magazines and request their field test reports on any models you are considering. These magazines also publish annual buyer's guides, which are helpful in revealing what is available. You should pay attention to the quality and variety of lenses and accessories available for different brands, but there is no need to buy top-of-the-line equipment to get good results. Remember the story of the photographer and the writer who met at a party. "I saw your exhibit at the gallery last week," said the writer. "You must have a very good camera!" To which the photographer replied, "Why, thank you. By the way, I really enjoyed your latest book. You must have an excellent word processor!"

Certain accessories are quite useful to the practical entomologist, while others have limited application in this field. Lenses are labeled by their focal length, which in turn determines the

Features to look for when selecting a camera

Mandatory
Single-lens reflex
Interchangeable lenses
Depth-of-field preview
Built-in light meter
Self-timer

Optional
Auto-exposure with manual override

Auto-focus with manual override
Motor drive
Dedicated TTL flash capability
Auto-rewind
Lock-up mirror
Interchangeable focusing screens
Built-in flash

perspective they yield. Most cameras are offered for sale with a standard 50mm lens, although you may wish to purchase the camera body alone and select your own lenses. A 50mm lens gives roughly the same perspective, or angle of view, as that of the human eye.

Lenses

Wide-angle lenses are those with a focal length of less than 50mm. These provide a wider field of view than seen by the human eye, and consequently the subjects appear smaller, although the difference is negligible until you get down to a focal length of 35mm or less. Wide-angle lenses are useful for taking photos of landscapes and habitats, and also for photographing large subjects

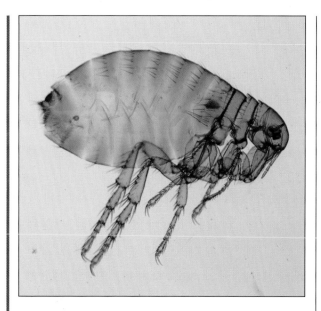

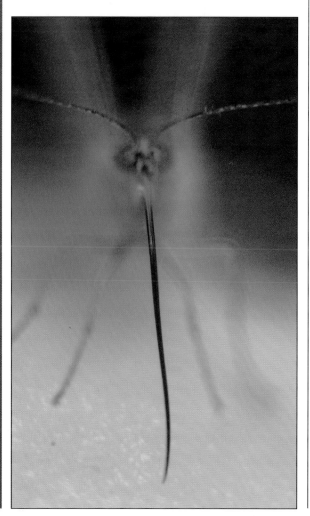

at a close distance. They have a limited application in photographing insects.

At the other extreme are telephoto lenses, those with focal lengths ranging from 50mm to 1,000mm or more. However, only the medium-length telephoto lenses, up to approximately 200mm, have any practical application in photographing insects, and then only when used with close-up accessories that will be discussed later. Telephoto lenses provide a narrower field of view and subjects appear larger than when seen by the unaided eye. They are used primarily when you cannot get as close to your subject as you wish, or in order to isolate the subject and remove a cluttered foreground.

High-quality zoom lenses, providing variable focal lengths, are extremely practical for field work. The entire range of focal lengths from about 24mm up to 210mm can be covered with just two lenses. This leaves you relatively unencumbered and able to carry other gear as well. Adjusting the focal length allows you to compose a photograph exactly as you want it without having to move the camera closer to or farther from the subject.

Macro accessories

Unlike most other fields of nature photography, close-up equipment is necessary for insect photography. Macro lenses can be used at their designated focal length, but also allow you to continuously focus down to very close distances, usually a matter of a few inches. They can thus do double duty, and many photographers prefer to purchase a 55mm macro lens instead of the

Top left Results like this can be achieved with some experimentation and a special camera attachment for a microscope. The insect can then be magnified many times, and photographed with a 35mm SLR.

Left This close-up of a butterfly proboscis was taken with a 50mm lens, three extension tubes and a tripod. Close-up equipment enables you to photograph in minute and fascinating detail, but it may take some practice before you become expert.

standard 50mm lens offered with their camera body. Macro lenses are available in a variety of focal lengths, and the shorter the focal length, the closer you must be to the subject in order to achieve maximum magnification. The best macro lenses give 1:1 reproduction, which means that the image on the frame of film is the same size as the actual subject.

Results equal or superior to those possible with macro lenses can be achieved with standard lenses by adding space between the camera body and the lens, usually by means of extension tubes or bellows. Extension tubes are less expensive and more practical to use, but offer somewhat less versatility in achieving a specific image size.

With the aid of an adapter, you can convert your ordinary lenses into high-quality macro versions by reversing them on the camera body. Similarly, you can purchase adapters to reverse one lens on top of another to achieve super-magnification. One drawback to both methods is that the aperture on the reversed lens must be set manually, negating any auto-exposure capability your camera may have. Also, by linking two lenses you will dramatically reduce the amount of light reaching the film, which means that you will almost certainly need to use a flash.

One close-up accessory that is a great convenience, if not a necessity, is a focusing rail. The range of distances from the camera that is in crisp focus, called *depth-of-field*, is reduced proportionately to the magnification being used. This means that when using the high magnification possible with some macro equipment, it is often difficult or impossible to hold a camera steady enough to keep what you want in focus. By mounting the camera on a focusing rail and securing that on a tripod, you can make fine adjustments in camera position.

Tripod

One of the most effective accessories you can use to improve the quality of your photographs is a sturdy tripod. Most blurred images result not from poor focusing but from camera movement. The steadiest person in the world cannot hold a camera as still as a mediocre tripod. Although a tripod can improve sharpness at any shutter speed, the universal rule of thumb is not to hold a camera when shooting at less than the reciprocal of the focal length of the lens. Put more simply, when

Top When using a large aperture (f2.8) it is important to focus carefully on your subject as the foreground will be slightly out of focus.

Above When using a small aperture (f.22) the depth of field increases and the overall focus is sharp.

Guidelines for purchasing a tripod

• Buy the best tripod you can afford, best in this case meaning sturdiest, not the fanciest. Before you buy, set up the tripod and lock everything in place, then wiggle, twist, and turn every part of it. Nothing should move – not at all.

• Select a tripod that allows you to shoot at or near ground level, but avoid those with reversible center columns. These do indeed permit you to shoot at ground level, but oblige you to work with a tripod leg between you and an upside-down camera, a nearly impossible task for most of us.

• Buy a tripod that reaches your eye level *without* the center column extended, which would effectively convert your tripod into a less stable three-legged monopod.

• Make sure the tripod and camera are light enough for you to carry them without undue strain. The strongest tripod in the world is completely useless if it has to be left at home because it is too heavy.

• Buy and use a cable release with your tripod. This will prevent camera movement caused by depressing the shutter button with your finger.

Above Here, a camera is mounted on a tripod, and the 35mm lens is separated from the camera body by bellows (a close-up accessory); the cable release trips the shutter without blurring the image.

using a 50mm lens, you should use a tripod or other steady support if your shutter speed is less than 1/50 of a second. Hand-holding a 200mm lens requires a shutter speed of at least 1/200 of a second, and so on.

When you need a tripod but can't use one because your subject doesn't stay in one place very long, a shoulder stock may solve the problem. This permits you to steady the camera against your shoulder and aim it like a rifle. The grip even has a trigger connected to a cable release that trips the shutter. A monopod is also useful for impromptu occasions.

Films

Available films vary nearly as much as camera models. What is right for you depends upon how you want to use your images, and the conditions under which you are shooting. Print films are developed into negatives, from which a print is made on light-sensitive photographic paper. Prints are easy to display and view, whether in an album, or framed on a wall, or otherwise. Black-and-white prints will produce an accurate representation of the form and texture of your subject, but they cannot portray the vast array of colors among insects. Color prints are more aesthetically pleasing and provide a more accurate representation of natural subjects than do black-and-white prints, but they are also the most expensive types of film to develop, thus limiting those on a tight budget.

Color slide films, also known as color reversal films, are the workhorse of professional nature photographers, especially those shooting for publication, since transparencies (and black-and-white prints, to a lesser extent) are used almost exclusively in the publishing business. Transparencies cannot be viewed easily without a projector, slide viewer, or a loupe and light table. Their major application, aside from publishing, is their use in illustrating lectures. Slide film is much less expensive to develop than color print film, and you can also have prints made from slides at a custom photo laboratory. Although it costs more per picture than print film, you can afford to experiment with different exposures and compositions and print only those you like.

The other major variable associated with films is light sensitivity, better known as film speed and denoted by an ASA (American Standards Association) number. The higher the ASA number, the faster, or more light sensitive, is the film. Slower films therefore require a slower shutter speed *or* a larger lens opening under a given light intensity than do faster films.

Greater light sensitivity is achieved by making the grains on the film emulsion larger. Photographs produced with faster films are consequently more grainy than those produced with slower films. In other words, the lower the ASA number, the sharper the image. For this reason, publishers generally will not consider using images produced on film rated higher than 100 ASA. High speed films are most useful under low light conditions, if you have chosen not to use a flash, or when you want to use a very fast shutter speed to freeze movement.

Exposure

A photograph records light. In order to achieve the correct exposure, the correct amount of light must reach the film. This is a function of the combined shutter speed and size of the lens opening, called the *aperture* or *f-stop*. Modern 35mm cameras have built-in light meters which measure the amount of light reaching the film. Most cameras can use this information to calculate the correct aperture for a given shutter speed, or vice versa, and some will calculate both.

Making a correct exposure is very much like trying to fill a bucket exactly to the top with water from a faucet. A slow trickle takes longer to fill the bucket than opening the faucet all the way. If too much water is added, the bucket overflows; with too little water, it is not as full as you wish. Correspondingly, the correct exposure of a photograph requires the use of a longer shutter speed with a small aperture than with a large one, and vice versa. Too much light results in overexposure with washed-out highlights, and too little results in an underexposed image.

Natural light vs. electronic flash

Whether or not you use an electronic flash will depend on several factors. If you want your photograph to look as natural as possible, try to use whatever light is available at the scene. Bright overhead sunlight will result in harsh, unappealing shadows, but on smaller subjects you can fill these in by using reflectors made by covering cardboard with crinkled aluminum foil. Shadows may also be softened by constructing a light-diffusing lean-to

Top Because the bees are not moving quickly a shutter speed of 1/4 second has proved suitable. At such slow speed a tripod is essential.

Above Photographs taken in strong sunlight often result in harsh shadows. Using a flash has softened the contrasts while highlighting the shiny bodies of the bees.

from translucent sheets of plastic, such as those used to cover fluorescent light fixtures.

Low natural light will require that you use a large aperture, a slow shutter speed, or both. Slower shutter speeds demand the use of a tripod to attain sharp focus, but a constantly moving subject will effectively prohibit the use of slow shutter speeds, even with a tripod. In cases like this, or when you wish to darken the background and emphasize the subject, electronic flash is the best choice. Using flash also allows you to fill in harsh shadows, freeze movement, or use a smaller aperture to

Top Fast films are useful in low light conditions where it is undesirable to use a flash. However, the image appears grainy, as in this 400ASA picture.

Above Slower films require more lighting but produce a sharper image, as in this 100ASA picture.

increase depth-of-field, a concept we shall discuss shortly.

The most aesthetic electronic illumination of a small subject is produced when the flash is positioned close to that subject and above the camera lens, producing soft shadows and duplicating the natural angle of illumination from the sun. To achieve consistent results when using lenses of different focal lengths, you will need to buy or make a bracket that allows you to position the flash over the same reference point, such as the end of the lens, every time. The flash itself need not be powerful since it is so close to the subject.

Another accessory, the ring flash, eliminates the need for a specialized bracket. A ring flash is circular and attaches to the front of your lens. While it produces consistent results, the even illumination that it produces tends to wash out shadows and may produce photos that lack a sense of depth. Also, any highlights reflected within the photograph will appear as small circles rather than natural-looking points of light.

Focus

Another function of aperture size is depth-of-field, the area within a certain range of distances from the camera which is in focus. Using a given lens, a smaller aperture will yield a greater depth-of-field than a larger aperture. Should you desire to emphasize your subject by focusing on it rather than on the background or foreground, you would select a larger aperture and a faster shutter speed. If you desire more depth-of-field, to portray a plant in its natural habitat for instance, you select a small aperture, but to get the correct exposure you must also employ a correspondingly slow shutter speed. The focal length of a lens also affects the depth-of-field. Shorter focal lengths yield greater depth-of-field than longer focal lengths. Most lenses have a scale that will show your depth-of-field in feet and meters from the camera, so you can calculate which parts will be in focus.

Recommended photographic equipment

Option 1	Option 2
35mm SLR camera body	35mm SLR camera body
55mm macro lens	28–70mm zoom lens
105mm macro lens	70–210mm zoom lens
200mm macro lens	Extension tubes/bellows
Extension tubes/bellows	Focusing rail
Focusing rail	Warming filters
Warming filters	Polarizing filters
Polarizing filters	Electronic flash
Electronic flash	Tripod
Tripod	Cable release
Cable release	Reflector
Reflector	Diffuser
Diffuser	

KEEPING A FIELD NOTEBOOK

Insect field guides and identification keys are wonderful tools that provide excellent background information. However, *you will learn and remember more about insects from your own detailed observations than you will from reading about them!* For this reason, a meticulously maintained field notebook is by far your most valuable learning tool. Notes about anything that strikes you as interesting or significant will serve you for the rest of your life.

A field notebook is not a diary. You need not make an entry every day, week, or at any regular intervals. To get the most from a field notebook, however, it should be part of the basic gear that you take along every time you go into the field, whether that be for collecting, photographing, or just taking a hike. If your interests include other subjects besides entomology, these also have a place in your field notebook. Basic information in each entry should include the date, departure and return times, weather conditions, temperature, and geographic locations and habitats visited.

Always record your observations on the spot; important details will be forgotten if you trust them to memory. Even a short note with a few key words will help jog your memory and allow you to fill in the details later. Also, regardless of your artistic abilities, always include many sketches with your notes. No matter how crude, sketches are great memory triggers that help you to visualize what you saw while making the sketch. Sketching also trains you to make precise observations, a skill critical to any naturalist. You may find that drawing only a portion of the organism helps to focus your observations and reveals greater detail.

Logbook

As a supplement to your field notebook, consider keeping a loose-leaf logbook at home, into which you can transcribe your notes, reorganize them, expand them, and co-relate them with your readings and past observations. Do this without fail after each excursion, and before you know it you will have amassed an impressive body of knowledge, some of which may not be found in field guides and, in fact, may never have been recorded.

Catalogue card file

In addition to, or in place of, a logbook, you may find it useful to start an index card file on the families and species that you have come to know. Use one card per discovery and record its name, date found, relevant information, and the page number of your field notebook on which you make reference to it. Number each card and use the corresponding number to identify photographs, specimens, and notebook entries. It is easier to have your notes on file in this manner than to search through notebooks for information.

Don't expect to learn everything about a particular species the first time you find it. Think of each encounter as a jigsaw puzzle with the pieces all askew. With each encounter, you have the opportunity to put a few more pieces into place. In time, an image begins to emerge, and the background of the picture also becomes more distinct as you learn more about how the organism interacts with its surroundings. Keeping a field notebook and a logbook or card file enables you to save and assemble all the pieces of the puzzle.

Typical Catalogue Card

Specimen number
Common name
Scientific name
Order
Date
Location
Habitat
Description

Notebook page number

Family

KEY TO INSECT ORDERS

The following key will give you a starting point from which to identify any unfamiliar insect by placing it in the correct order. Begin with number one and decide which of the two options, a or b, fits the insect in question. Whichever one you choose will either reveal the order or direct you to another part of the key, and so on, until the order is disclosed. For best results, it is important to read each question carefully and examine the insect meticulously. Some orders have members that are both winged and wingless, and hence are listed twice in the key.

1 Does the insect have wings?
a. Yes .... Go to #2
b. No .... Go to #17

2 How many pairs of wings does the insect have?
a. One .... Diptera
b. Two .... Go to #3

3 Do the two pairs of wings differ greatly in structure, the first pair being thick and hard or leathery?
a. Yes .... Go to #4
b. No .... Go to #7

4 Is the first pair of wings rigid, and do they meet in a straight line down the middle of the back?
a. Yes .... Go to #5
b. No .... Go to #6

5 Is there a pair of prominent pincerlike cerci at the tip of the abdomen?
a. Yes .... Dermaptera
b. No .... Coleoptera

6 Does the insect have:
a. chewing mouthparts, front wings leathery and heavily veined, and hind wings folded like a fan? .... Orthoptera
b. sucking mouthparts and front wings leathery at the base, membranous and overlapping at the tip? .... Hemiptera

7 Are the mouthparts a coiled tube and the wings covered with scales?
a. Yes .... Lepidoptera
b. No .... Go to #8

8 Are the wings rooflike, sloping downward and outward from the middle of the back?
a. Yes .... Homoptera
b. No .... Go to #9

9 Is the insect slender and mothlike, with long, slender antennae and wings that are widest past the middle?
a. Yes .... Trichoptera
b. No .... Go to #10

10 Do the wings have few or no cross veins?
a. Yes .... Go to #11
b. No .... Go to #12

11 Does the insect have chewing mouthparts and hind wings somewhat smaller than the front wings?
a. Yes .... Hymenoptera
b. No .... Thysanoptera

12 Are there two or three long, slender, tail-like appendages on the tip of the abdomen?
a. Yes .... Ephemeroptera
b. No .... Go to #13

13 Does the head have an elongated trunklike beak with chewing mouthparts at its tip?
a. Yes .... Mecoptera
b. No .... Go to #14

14 Does the insect have inconspicuous antennae, long narrow wings, and a long slender abdomen?
a. Yes .... Odonata
b. No .... Go to #15

15 Does the insect have two short cerci on the tip of its abdomen and front wings narrower than the rear wings?
a. Yes .... Plecoptera
b. No .... Go to #16

16 Do the tarsi each have 5 segments?
a. Yes .... Neuroptera
b. No .... Isoptera

17 Is the insect antlike, with a narrow waist?
a. Yes .... Hymenoptera
b. No .... Go to #18

18 Is the insect antlike, but with a wide waist?
a. Yes .... Isoptera
b. No .... Go to #19

19 Is the insect small and flattened, with chewing mouthparts and a head about as wide as its body?
a. Yes .... Go to #20
b. No .... Go to #21

20 Are the antennae long, and composed of many segments?
a. Yes .... Psocoptera
b. No .... Mallophaga

21 Is the insect's body soft and rounded, with two short tubes protruding from the abdomen, and with a small head?
a. Yes .... Homoptera
b. No .... Go to #22

22 Is the insect very small, with a vertically flattened body, a hooklike claw on each leg, and sucking mouthparts?
a. Yes .... Anoplura
b. No .... Go to #23

23 Is the insect very small and narrow (flattened laterally) with sucking mouthparts?
a. Yes .... Siphonaptera
b. No .... Go to #24

24 Is the insect:
a. delicate with chewing mouthparts and threadlike "tails" and antennae? Thysanura
b. very small with a springlike lever folded under its abdomen which it uses for leaping? .... Collembola

DRAGONFLIES & DAMSELFLIES
ORDER ODONATA

INTRODUCTION

Dragonflies and damselflies, large and often brightly colored, are arguably the most accomplished aerialists in the animal kingdom. Dragonflies, in fact, are not only remarkably swift and agile, able to travel at up to 35 miles per hour and reverse direction in midair within one body length, but they can also hover with ease and fly backward like tiny helicopters. Their wings are reinforced with deep corrugations at the base, which accounts in part for their strong flight.

Most of the stronger insect fliers have evolved means to link their fore wings and hind wings in flight so that each side functions as one flight surface. Dragonflies and damselflies, in contrast, move the two pairs independently, timing the stroke of the hind pair so that they meet the oncoming air before it has been disturbed by the front pair; physicists would describe them as operating in *antiphase*. Even more remarkable, dragonflies have evolved a technique that endows them with tremendous power for their size. Researchers studying dragonflies in a wind tunnel have found that dragonflies twist their wings on the downstroke, creating miniature whirlwinds that move the air much faster over the upper wing surface, reducing the air pressure there and greatly increasing lift. These findings are deemed to have so much potential value in aviation that

the United States Navy and Air Force have contributed to the funding of such research.

Damselflies and dragonflies catch their prey on the wing. Their long, slender legs, each with a row of stiff bristles on either side, are held slanting forward in flight to form a basket that scoops smaller flying insects out of the air and traps them until the jaws can grasp them. Their sharp chewing mouthparts enable them to cut up their prey into bite-sized pieces. They are quite beneficial to humans in that they consume huge numbers of mosquitoes and black flies.

The success of these "dragons of the air" as aerial hunters is due largely to their acute vision. Their huge compound eyes, each composed of 10,000–30,000 individual facets, are supremely adapted for detecting motion, and their heads have an unusually large range of motion for insects, enabling them to be cocked at different angles and perhaps improving their depth perception from different perspectives.

Dragonflies and damselflies copulate in the air, and you can observe these mating flights over any body of fresh water in mid-summer. Each male curls the end of his abdomen forward to place a packet of sperm in a cavity underneath his second abdominal segment, then grasps a mate by her neck with clasping cerci at the tip of his

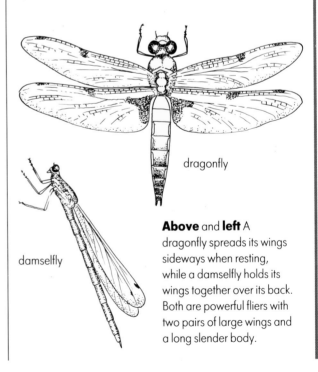

dragonfly

damselfly

Above and **left** A dragonfly spreads its wings sideways when resting, while a damselfly holds its wings together over its back. Both are powerful fliers with two pairs of large wings and a long slender body.

What you can do

Dragonfly watching

Though it will never attain the popularity of bird watching, observing dragonflies and damselflies has its own rewards. By employing slow, steady movements on land or using a drifting canoe or rowboat, one can approach these wary insects quite closely and observe territoriality, mating, egg-laying, predation, and even metamorphosis, all within a relatively small area.

Above Freshly-emerged, an adult darner dragonfly (*Aeshna juncea*) dries its expanded wings. Its body and wings are still a downy light color that darkens as the insect dries.

abdomen. In flight, she curls the tip of her abdomen forward to retrieve the sperm capsule. She later deposits her fertilized eggs in or near the water.

Naiads

Odonatans undergo hemimetabolous development, but with an interesting twist. The immature insects, called *naiads*, are aquatic and bear little resemblance to the adults. They breathe by means of tracheal gills lining the rectum, and damselfly naiads also have three long tail-like gill plates extending behind the abdomen.

Naiads of this order are among the chief aquatic insect predators. They lurk stealthily among vegetation or lie quietly concealed by silt and debris, but dragonfly naiads can dart forth with amazing speed, thanks to their jet propulsion system that shoots a powerful stream of water from the tip of the abdomen. Another weapon at their disposal is an elongated hinged lower lip, or labium, with a pair of grasping jaws at the tip. While usually folded under the head with the scoop-shaped tip masking the face, it can shoot forward with lightning speed to snatch prey an impressive distance ahead.

When they are ready to become adults, damselfly and dragonfly naiads crawl up plant stems, sticks, dock pilings, or any other structure protruding from the water. Here, each undergoes its last molt and flies away, leaving the cast skin, or *exuvium*, still clinging to its perch, a common summer sight around lakes and ponds.

Along with mayflies, dragonflies and damselflies are the most ancient of flying insects. Their ancestors, some of which developed wingspans of up to 30 inches first appeared during the Carboniferous Period, between 350 and 280 million years ago. Today there are about 5,500 species of dragonflies and damselflies worldwide, with approximately 450 species in 11 North American families. They are usually found over water, where males will defend territories against rival males of the same species. Members of this order are easily recognized by their bulging compound eyes, heavily-veined wings, and long, slender abdomens. They cannot fold their wings flat against their back, as can many other orders. Instead, dragonflies rest with their wings to the sides, perpendicular to the thorax, and damselflies hold their wings vertically and rearward.

69

DRAGONFLIES

Darners – Family Aeshnidae

Darners are the largest and swiftest of North American dragonflies. Adults range in size from 50–120mm in length and may have wingspans up to 150mm. Brilliant green, blue, or brown, with clear wings, they are easy to recognize, being the only family of dragonflies whose compound eyes meet along the top of the head. Darners are also the dragonflies most likely to be found fair distances from water, although they usually haunt ponds and slow streams. They are named for their resemblance in flight to darning needles.

Clubtails – Family Gomphidae

Named for their swollen, terminal abdominal segments, clubtails favor streams over ponds. They soar and hover much less than other dragonfly families, preferring instead to rest on logs, stones, and leaves. They catch prey while darting from one resting place to another. The bodies of adult clubtails are 31–90mm long and darkly-colored with yellow or green stripes. Identifying features include their widely separated eyes and spreading legs.

Biddies – Family Cordulegastridae

A small family, biddies haunt small woodland streams almost exclusively, cruising slowly about 12 inches over the surface. Adults are 45–88mm long and smoky brown, with yellowish markings and a hairy appearance. Their compound eyes meet or are slightly separated at a single point only.

Green-eyed Skimmers – Family Corduliidae

Adults of this family are 28–78mm long and usually black or metallic, with bright green eyes, although some are dark brown with yellow markings. The hind margin of each compound eye is distinctly lobed. Green-eyed skimmers favor woodland swamps and ponds.

Belted Skimmers and River Skimmers – Family Macromiidae

Belted skimmers are brown with lighter markings. They are uncommon, but usually occur along the marshy margins of ponds. River skimmers, denizens of rivers and slow streams, are blackish

Above The hairy thorax of this golden-ringed dragonfly (*Cordulegaster boltoni*) is obvious, indicating that it is freshly-emerged from the naiad. The naiads breed in muddy moorland streams or in boggy areas.

Above The four-spotted chaser dragonfly (*Libellula quadriaculata*) inhabits boggy pools and ponds. Males may be territorial, defending their own patches.

Left The wide and comparatively short body of the broad-bodied chaser dragonfly (*Libellula depressa*) is typical of this group of dragonflies, and in this female the orange markings are highlighted on the base of the wing veins. The articulation joints are clearly visible on the wing bases.

Above The tail of this female club-tailed dragonfly (*Gomphus vulgatissimus*) is a key identification feature. A member of the Gomphidae family, it has a characteristic small, heart-shaped head. The naiads breed in slow-flowing streams and rivers.

with yellow markings. Both range from 56–91mm in length.

Common Skimmers – Family Libellulidae
Common skimmers are the largest family in this order and are represented by the most common dragonfly species. They have brightly colored bodies, 18–75mm long, that are noticeably shorter than the wingspan. The males and females of some species are colored differently. Most dragonflies whose wings have colored bands or spots belong to the Libellulidae. They're fond of perching on plants along the margins of ponds, lakes, marshes, and quiet streams. The flight of common skimmers is swift, sometimes interrupted by intervals of hovering.

What you can do

Underwater jets
To observe the unique propulsion system of dragonfly naiads, set up an aquarium with about 8cm of muddy water and allow a thin layer of pond silt to settle on the bottom. Introduce one or more immature dragonflies; prod them gently with a blunt stick to get them moving, and you will see a stream of water shoot from the tip of the abdomen and stir up silt as the dragonfly darts forward. This is primarily an escape mechanism, as the naiad relies upon stealth to capture prey.

71

DAMSELFLIES

Broad-winged Damselflies – Family Calopterygidae

A striking family, the broad-winged damselflies often display bright metallic bodies, 25–51mm in length, and wings that are dark or have brightly colored spots. Though their wings narrow at the base, they are not stalked, as are those of other damselfly families, hence their name. They tend to frequent woodland streams, often perching on overhanging vegetation.

Narrow-winged Damselflies – Family Coenagrionidae

Narrow-winged damselflies include more species than any other damselfly family. These brightly colored insects, measuring 25–50mm in length, prefer quiet waters, such as ponds, marshes, and swamps. Colors and patterns usually differ bet-ween the sexes, with males being the more colorful of the two. Most have clear wings, extremely narrow or stalked at the base, that are held together vertically over the body when at rest

Spread-winged Damselflies – Family Lestidae

The clear wings of this family are also stalked, but unlike those of the narrow-winged damselflies, they are held partly spread over the body at rest, instead of vertically together. Ranging in length from 32mm to 62mm, members of this family are most often found around the margins of pond, swamps, and marshes.

Below The banded demoiselle (*Calopteryx splendens*) stands delicately on a leaf with its tarsi and claws while showing off its bristly legs. This male is ready to attack.

Right This side view of a large red damselfly (*Pyrrhosoma nymphula*) reveals its large thorax, which houses the powerful flight muscles, as well as the long abdomen.

What you can do

Tips for collecting

Capture techniques
aerial net
Killing method
ethyl acetate killing jar
Preservation techniques
pin on spreading board

2 Place dragonflies and damselflies in triangular paper envelopes for protection, with their wings folded above the body.

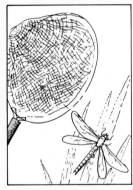

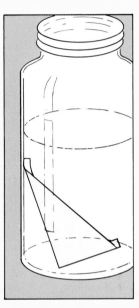

1 Dragonflies are so adept at evading other aerial predators that obtaining specimens can prove quite frustrating, even with a net. The secret is to remember that resting dragonflies take off at a 45 degree angle forward when alarmed, so wait for your specimen to land; approach slowly; place the net in the projected path of escape, and startle the dragonfly.

3 The bright colors on many specimens tend to fade, but this can be minimized by rapidly drying them in the sun, an oven, or under a lamp, after they have been pinned. You can also preserve their colors by keeping them in the paper triangles and sealing these in a jar of acetone for 24 hours, then allowing them to air dry. Acetone is flammable, so use caution with this method.

GRASSHOPPERS, KATYDIDS, & CRICKETS
ORDER ORTHOPTERA

INTRODUCTION

The Order Orthoptera includes most of the champion jumpers of the insect world. Proportionately, were we as good, we could clear over 300 feet in the high jump and 500 feet in the broad jump! Jumping high, far, and fast is their primary defense against predators, so those that leap best also survive longer and reproduce more successfully, creating even better jumpers in future generations. Characteristically, most families have greatly swollen femurs on their hind legs, housing powerful jumping muscles. Many species in this order also rely heavily on camouflage for protection.

There are about 1,200 North American species of these medium-to-large insects in 10 families, and about 20,000 species worldwide. Orthopterans usually have narrow, leathery forewings and broad, membranous hind wings that fold accordion-style under the forewings when at rest.

Orthopterans undergo hemimetabolous development. Both the nymphs and the adults have chewing mouthparts and are mainly vegetarians, although some cannibalism has been observed when high populations deplete the available food supply. Some are omnivorous, and others, especially those that live underground or within human dwellings, are scavengers, eating decaying plant and animal matter.

Sound is very important in the courtship of most orthopterans. Consequently, they have some of the most well-developed tympanic organs among insects. These sensitive structures are located either on the tibia of each front leg or on each side of the first abdominal segment.

Depending upon the family, their songs, called *stridulations*, are produced either by rubbing their spiny hind femurs against a projecting vein on the outside of the forewing (*tegmen*) or by rubbing a hard scraper on one fore wing against a file-like vein on the other, much like one would draw a bow over violin strings. Each song is unique to an individual species, so there is no mistake as to who is calling. It is the males that sing, and the females, recognizing the song of their species, seek out the serenader.

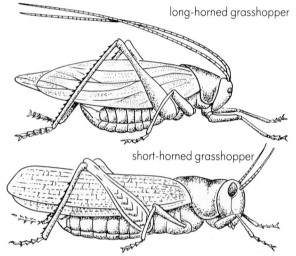

long-horned grasshopper

short-horned grasshopper

Above Crickets and grasshoppers are noted for jumping and singing. Most families have swollen femurs housing powerful jumping muscles, and some of the best developed tympanic organs in the insect world.

What you can do

Collecting tips for the Order Orthoptera

Capture techniques
aerial net
sweep net
hand collecting
beating tray
Killing method
ethyl acetate killing jar

Preservation techniques
pin (Specimens with a flash pattern on the hind wing may be mounted with one wing spread.)

Left The well-camouflaged Egyptian grasshopper (*Anacridium aegyptium*), from North Africa and southern Europe, has a lugubrious appearance. Its compound eyes are banded like the pattern on the elytra of a Colorado potato beetle.

Above Resting on a tropical zinnia, a bush cricket of the family Tettigonidae (*Tylopsis lilifolia*) has enormous antennae that make it aware of its environment, particularly of scents and touch stimuli.

GRASSHOPPERS & KATYDIDS

Pygmy Grasshoppers and Grouse Locusts – Family Tetrigidae

Tetrigids may occupy many habitats, but they are most often found in damp areas such as stream borders. Though terrestrial, they can swim if necessary. Adults overwinter in sheltered areas and are most often found in spring and early summer. The most notable feature of members of this family is the pronotum, which is long, tapered, and extends backward extensively and part of the abdomen. The forewings are small and may be hidden under the pronotum, but the hind wings are usually well-formed. These are small to medium insects.

Long-horned Grasshoppers – Family Tettigoniidae

Noted for their long, slender antennae, which are as long or longer than the body, long-horned grasshoppers are medium-to-large (14–75mm) insects, and include the well-known katydids. They vary from brown to bright green in color, making them difficult to distinguish from the vegetation they inhabit. Their green colors are absorbed from the chlorophyll in the foliage they eat; in fact, many effectively mimic leaves or other plant parts as a defense against predators. Most prefer the dense leafy cover of trees and shrubs, but some can be found in wet grassy meadows or the borders of streams.

Aside from their antennae, long-horned grasshoppers are also distinguished by the tympanum at the base of each front tibia, by their 4-segmented

Above and **left**: A natural wonder – the mountain grasshopper (*Acripela reticulata*) from Australia. The insect relies on surprise to scare its enemies, showing off its bright red and blue abdomen and the shiny undersides of its forewings in a defensive display designed to deter predators.

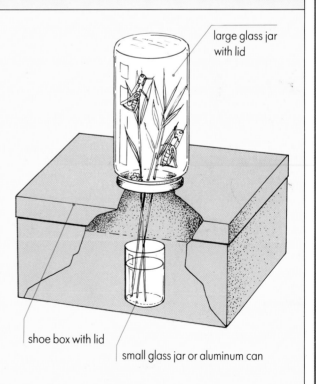

tarsi (feet), and also by the swordlike ovipositors used by females to slit plant tissue in order to lay their eggs. Males of this family are among the best songsters of all insects, especially the katydids, whose rasping songs are variations on the tempo of their name, "katy-DID, katy-DIDN'T," and so on. Long-horned grasshoppers overwinter as eggs.

Short-horned Grasshoppers – Family Acrididae

When grasshoppers are mentioned, most people will immediately think of the members of this very common family. These medium-to-large insects (12–80mm long) are named for their short, stout, horn-shaped antennae, which are usually less than half as long as the body. Another notable trait is the location of their large tympanic organs on the sides of the first abdominal segment. Males make a low buzzing song by rubbing the rough surface of the hind wing against the forewing.

Short-horned grasshoppers may readily be found in all terrestrial habitats up to about 14,000 feet in elevation, but are most common in dry grasslands and deserts. They include the numerous species referred to as "locusts." Many attack crop plants. Females exhibit short, stout ovipositors, which they use to lay masses of between 8 and 25 eggs below the ground's surface. Most short-horned grasshoppers overwinter in the egg stage, and do not hatch until summer.

Members of one subfamily of Acrididae, the band-winged grasshoppers, employ flash display, bold colors or patterns on their hind wings that burst into view upon takeoff and startle potential predators. As the insect lands, it quickly stows each boldly-marked hind wing under a dull tegmen and effectively disappears as it drops quickly against a similar patterned background. Any bird or mammal in pursuit usually continues to follow the original flight path after the conspicuous wings have folded, overshooting the motionless insect on the ground.

Spur-throated grasshoppers, another subfamily, includes those species most damaging to crops. Some species are migratory and may form swarms that number in the billions.

What you can do

A grasshopper zoo

Equipment

large glass jar with lid
small jar or aluminum can
shoe box
aerial net or sweep net
electric drill (or a hammer and a large nail)

Grasshoppers and katydids are interesting to watch and easy to keep in captivity. They eat plant foliage, and so should be provided with a few leafy plant stems from the area where they were captured, and a bit of damp sponge for humidity.

Turn the lid of the large jar upside down, and glue it to the top of the shoe box lid. After it dries, drill or punch several holes through the lid of the jar, making them large enough to accept medium-sized plant stems but small enough to prevent escape. Next, riddle the shoe box with holes for ventilation. Fill the can or small jar about three-quarters full with water and place it in the shoe box so the ends of the plant stems will rest in it when closed. Place your specimens in the jar; thread 2 or 3 plant stems through the lid; screw the lid onto the large jar; invert the jar and lid on top of the shoe box, and your zoo is finished.

large glass jar with lid

shoe box with lid

small glass jar or aluminum can

CRICKETS

True Crickets – Family Gryllidae

Shorter fore wings, 3-segmented tarsi (feet), and a broad and somewhat flattened body are the principal features distinguishing these medium-sized (9–25mm long) insects from long-horned grasshoppers. They are equipped with long feeler-like cerci on the tip of the abdomen, and females use needlelike ovipositors to lay eggs singly, either in the ground or on plant tissue. Like the long-horns, crickets also have stridulating organs at the base of their fore wings, but the shorter wings of crickets produce a higher-pitched and more musical song. A centuries-old practice in Oriental households is to keep caged crickets in order to enjoy their song.

Cave Crickets and Camel Crickets – Family Gryllacrididae

The members of this family are wingless, nocturnal, and inhabit moist locations under rocks and logs and in caves, trees, and the burrows of other animals. Each of the hind legs includes a stout femur and a long, slender, many-spined tibia. Many common species have a humpbacked appearance, and are gray or tan in color, with a length of between 10mm and 50mm. Few produce sound, the majority having no tympanic organs. Secluded and nocturnal, their sight is not nearly as keen as that of other orthopterans. Instead, they rely upon their very long and supple antennae to feel their way around. They are readily caught in pitfall traps baited with molasses and placed near wooded areas.

Mole Crickets – Family Gryllotalpidae

True to their namesake, mole crickets have scoop-like fore legs, modified for burrowing. Unlike most other orthopterans, their hind legs are not well-developed for jumping. They prefer moist sand and mud, particularly along streams and ponds. Dense coverings of hairlike setae keep damp soil from clinging to their bodies. Not surprisingly, eggs are laid underground, and both the nymphs and larvae feed on plant roots. Despite their subterranean lifestyle, mole crickets can fly well.

Above Because it spends most of its time underground, the mole cricket (*Gryllotalpa* sp.) has a thickened thorax and its first pair of legs are especially thickened for burrowing.

Right A common way of startling predators is through a hind-wing display, such as this performance by a bush-cricket (*Neobarrettia vannifera*).

Left The wood cricket (*Nemobius sylvestris*) is a lively resident of woodlands in south-western Europe, and shares its habitat with the stick insect. Of very nervous disposition, it seeks shelter when disturbed or jumps using its strong back pair of legs. Its long antennae are used to seek for food as well as keeping a general sensory watch on its immediate environment.

What you can do

A cricket thermometer

Equipment
stopwatch, watch with second hand, or digital watch
accurate air thermometer
pencil
field notebook
small flashlight or headlamp

As they are cold-blooded animals, the body temperatures of insects are rigidly governed by the temperature of the surrounding air. Within normal temperature ranges, their rate of activity and metabolism increases and decreases with the air temperature.

Approximate formulas have been worked out to determine the air temperature by the song of the snowy tree cricket (*Oecanthus fultoni*), which is common over most of North America. If T is the air temperature in degrees Fahrenheit and n is the number of snowy tree cricket chirps per minute, then $T = (n-40)/4 + 50$. In other words, to arrive at the temperature in degrees Fahrenheit, count the number of chirps per minute, subtract 40, divide the difference by 4, and add 50. A shortcut to this formula is to count the number of chirps in 14 seconds and add 40.

To arrive at the air temperature in degrees Celsius, use $T = 5/9 (n + 8)$, where n is the number of chirps in 14 seconds. In other words, determine the number of chirps in 14 seconds, add 8, multiply the sum by 5, and divide that product by 9. Who says math is no fun?

The possible variations of this are numerous. Can you correlate the air temperature to the songs of other crickets, grasshoppers, or katydids common in your area? At what temperature do they stop singing? Can the same relationship be demonstrated with other insect behavior (the flashing of fireflies, for instance?)

THE TRUE BUGS
ORDER HEMIPTERA

SELECTED FAMILIES

The word "bug" has evolved into a very general term, variously used to describe any and all insects as well as many other multi-legged creatures, such as spiders, scorpions, ticks, and mites. Technically speaking, though, the only true bugs are the members of the Order Hemiptera.

Hemiptera means "half wing," referring to the characteristic forewings of this order, called *hemelytra*, which are thick and leathery near the base and membranous near the tip, where they overlap. These wings fold flat over the insect's back covering the hind wings, which are the principal flying apparatus and are uniformly membranous and somewhat shorter than the forewings. Between the folded wings is a prominent triangular sclerite called the *scutellum*, the modified dorsal portion of the metathorax.

A third identifying feature of hemipterans is their piercing-sucking mouthparts, housed in a beak-shaped *rostrum*, or labium, that originates far forward on the head and is folded underneath it, pointing backward between the front legs. The rostrum swings out from its resting position as the bug prepares to feed. Precise directional control over this feature has allowed some hemipterans to exploit food sources other than plants, to which the closely related Order Homoptera is restricted.

True bugs may be either carnivorous, preying upon other insects and sometimes small vertebrates, or herbivorous. The mandibles, the outer pair of four thread-like mouthparts housed in the rostrum, have sharp teeth that are used to pierce the outer tissue of the victim. The inner pair of mouthparts, the maxillae, fit snugly together to form a salivary canal and a food canal. Saliva containing digestive enzymes is pumped down through the salivary canal, and the partially digested fluids are sucked up through the food canal. Many of the predators are beneficial to humans, consuming other insects regarded as pests. Predatory bugs frequently have *raptorial* front legs that are short, stout, and adapted for grasping and holding prey as they feed.

For defense, most true bugs have special stink glands that secrete a substance with a disagreeable odor in order to repulse enemies. Among nymphs, these glands are located on the back of the abdomen, while adults utilize glands located on the sides or underside of the thorax. Many advertise this capability with bright colors and bold patterns, recognized by animals that have had an unpleasant encounter with that species before. Others rely on cryptic coloration and employ their stink glands only as a secondary defense. Some true bugs can also deliver a painful bite when handled.

True bugs undergo hemimetabolous development with five developmental stages. Apart from sometimes varying in color and having wings that are reduced or absent, nymphs strongly resemble adults of their species. They develop from eggs that are most often attached to the surfaces of plants or implanted within their tissues. Hemipterans are widely distributed in most habitats, and

true bug

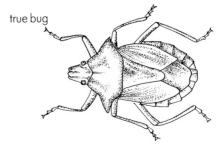

Above Although the term bug is used to describe many kinds of insects, technically speaking only members of the Hemiptera family are known as true bugs. There are over 23,000 known species world wide.

Far left Most aquatic hemiptera possess piercing and sucking mouthparts, but the lesser water boatman (*Corixa punctata*) has reduced mouthparts that are only suitable for eating algae and small animals; it also swims using enlarged and "feathery" legs.

Above left The water scorpion (*Nepa cinerea*) has a "tail" like that of a terrestrial scorpion, which it uses as a siphon for breathing through the water film while it hangs upside down, and lying in wait for its underwater prey.

Left Assassin bugs (reduvids) are true to their name. This one (*Apiomerus falviventris*) from Mexico has caught a bee and is sucking its juices with the same piercing mouthparts that it used to kill its prey.

although most are terrestrial, some very common species are aquatic, either with sprawling legs that distribute their weight over the surface of the water, or with oar-like legs that propel them over or through the water.

Hemipterans are best collected by sweeping and beating vegetation, looking under bark, or using a Berlese funnel or a light trap. Worldwide, there are more than 23,000 species of true bugs in over 60 families, with about 4,500 species in 44 North American families. We will be able to discuss just a few of the more significant families in this chapter.

Water Boatmen – Family Corixidae

Water boatmen are very common aquatic insects and the favorite food of many fish. They are easy to recognize by their elongated oval shape, 3–13mm in length, and by their broad head and eyes, which overlap the anterior end of the prothorax. Their dorsal side is flat, and finely barred with dark lateral lines that produce a mottled camouflage.

Water boatmen swim in a rapid, erratic fashion with oar-like strokes of their middle and hind legs, but most of their time is spent clinging to underwater vegetation, usually in shallow water near the edge of a pond or lake. Periodically, they visit the surface of the water to trap an air bubble, which they wrap closely around their bodies and under their wings, like a silvery envelope, in order to breathe. Their beaks (reduced mouth parts) are short and broad, and useless for piercing and sucking. Instead, they feed on algae, larvae, and other food particles scraped from underwater surfaces or strained from the water by the scoop-shaped tarsi of the short forelegs.

Backswimmers – Family Notonectidae

Most free-swimming aquatic animals have dark backs and light bellies, rendering them less visible against the light sky or dark depths. Backswimmers, true to their name, swim and float on their backs, and so their color pattern is reversed to provide the same protection against predators. Their fringed hind legs, much longer than their middle and front legs, are used as oars, and their keel-shaped back provides directional stability.

Backswimmers spend much time floating and resting at the surface of ponds and lakes, with head and body angled downward. They prey upon aquatic insects, terrestrial insects trapped in the surface film, and even small minnows and tadpoles. Members of this family trap bubbles of air in fringed pockets on the ventral surface, and can carry enough to survive up to six hours of inactivity underwater. Slightly longer than water boatmen (5–16mm long), backswimmers can also be distinguished from the former by their more streamlined shape, lack of dark cross-lines on the dorsal side, and lack of scoop-shaped tarsi on the forelegs.

Waterscorpions – Family Nepidae

Waterscorpions are easily recognized by their combination of scissor-like raptorial forelegs, sprawling middle and hind legs, and two very long abdominal appendages. When held together, these two filaments form a snorkel through which the insect breathes while clinging upside down to aquatic plants.

Most waterscorpions are slender and long – about 20–43mm in length – but those of the genera *Nepa* and *Curicta* resemble giant water bugs except for their tail-like breathing tube and lack of flattened hind legs. All are poor swimmers, and hence must remain motionless while awaiting the approach of unsuspecting prey.

Water Striders – Family Gerridae

Who among us does not have a childhood memory of idly staring at the edge of a pond or a quiet eddy in a stream and watching water striders skating about its surface with ease? They exploit the phenomenon known as surface tension, that property of water by which molecules at the surface cling more tenaciously to one another than they do elsewhere. This creates a film upon which lightweight objects can rest without breaking through. The same principle can be seen when water is

---

What you can do

Messing with Mother Nature

Have you ever been annoyed by the way in which impudent water striders taunt you, coming temptingly close, only to dash away before you can catch one? Here's how to get even! When you finally manage to capture one or more, place them in a container partly filled with clean, fresh water. Notice the dimples made by their feet as they depress the surface but do not break through.

Now, place a drop of dish detergent or light household oil on the water and watch your captives flounder. Both of these agents break the surface tension of water. You may wish to repeat this experiment with other substances to find out which have the same effect. Be sure to rescue your water striders before they drown, and discard the water away from the stream or pond.

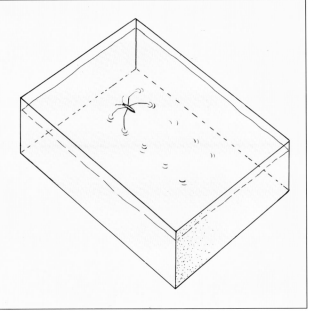

Above A small colony of the birch shieldbug (*Elasmostethus interstinctus*), from the family Acanthosomidae, contains one adult with fully developed wings, and four nymphs with developing wing buds.

Center In at the kill – water striders (*Gerris lacustris*) drink the blood of a dead insect (the large red damselfly, *Pyrrhosoma nymphula*) by piercing the hardened exoskeleton with their tough mouthparts. Water striders move rapidly over the surface film of the water with the aid of their long legs.

Left The brightly-colored *Lygaeus equestris* (family Lygaediae) is a common insect in southern Europe. It uses its warning colors as defense, and thus has no need to seek shelter as it sucks the juices from plants.

poured slowly into a glass until the level is higher than the lip of the glass itself. In the case of water striders, waterproof hairs on the undersides of their tarsi, combined with sprawling middle and hind legs that distribute their weight over a large area, enable them to walk on water.

Water striders prey upon the rain of insects that constantly fall onto any body of water in warmer months. Their struggle to escape the water sends ripples radiating across the surface, and the water striders use these as a form of homing signal. They use their raptorial front legs strictly for capturing prey, while darting across the water on their middle and hind legs. Water striders are generally slender and dark brown or black, their bodies measuring between 10mm and 25mm in length. This family includes the only true marine insects: certain species of the genus *Halobates*, found on warm subtropical seas.

Seed Bugs – Family Lygaeidae

Most members of this family, the second largest in the order, feed on seeds by injecting saliva and sucking out the partly digested contents. A few feed on other green plant tissues or sap, or prey upon other insects. They vary in appearance, with oval or elongated bodies from 3mm to 18mm long. Most are brownish, but some are boldly patterned with bright colors. They can be distinguished from members of most similar families by the presence of simple eyes and by the four or five prominent veins in the membrane of each forewing.

Leaf-footed Bugs – Family Coreidae

The odd name of leaf-footed bugs was derived from the unusual leaf-like expansions on the hind tibia of some species. Many also have enlarged hind femurs bearing prominent spines. These fairly large bugs, measuring from 7mm to 40mm long, can be differentiated from the seed bugs primarily by size and by the numerous parallel veins in the membrane of the forewing. Most leaf-footed bugs are plant eaters, and all can emit a foul odor when disturbed. Members of this family are also commonly known as squash bugs.

Stink Bugs – Family Pentatomidae

It takes little imagination to guess how these very common bugs got their name. In self-defense they will emit copious quantities of an offensive liquid from their thoracic glands to repel enemies. They

are generally between 6mm and 20mm in length and display a very prominent scutellum. The basal segment of their 5-segmented antennae is considerably thicker than the second. The many branching veins in the membrane of each forewing constitute another identifying feature. Most stink bugs feed on plant sap, but a few prey upon insect larvae.

Bed Bugs – Family Cimicidae

Though not a large family, bed bugs are significant in that they include a worldwide pest of humans, *Cimex lectularius*. They are often reddish-brown in color, measure less than 7mm in length, and are very flat and broadly oval. Their wings are reduced to tiny vestigial appendages that are barely visible. Bed bugs are nocturnal parasites that feed on the blood of mammals and birds, but they are not known to transmit any human diseases. By day, they hide in small cracks and crevices. Amazingly, adult bed bugs can survive up to 15 months without food.

Opposite top Typical of the family Coreidae, this leaf-footed bug (*Narnia inornata*) has enlarged hind tibiae, which help in the overall disguise of the insect on plants, and in rapid locomotion. Here, it is drinking juices from the fruit of the cactus *Pilosocereus sartorianus* in Mexico.

Opposite below The striking jester bug (*Graphosoma italicum*) – a European stink bug – has a matt surface to its striped body.

Above A mating pair of shield bugs (*Eurydema* sp.) from Corfu, Greece, display bright warning colors. They belong to the family of stink bugs (Pentatomidae) and emit deterrent odors if disturbed or handled. These common insects are often found in gardens.

What you can do

Collecting tips for true bugs

Capture techniques	Killing methods
hand collecting	ethyl acetate killing jar
aspirator	immersion in 70 per cent
sweep net	alcohol
beating tray	**Preservation techniques**
light trap	
	pin
	point

Plant Bugs – Family Miridae

Plant bugs constitute the largest family in the order Hemiptera, with some 1,700 North American species and roughly 6,000 species worldwide. Most feed on plant juices, and many are significant pests, but a few prey upon other insects. They are somewhat soft-bodied and often brightly colored. This family is marked by the presence of a *cuneus*, a distinct division of the wing along the anterior edge between the membrane and the basal section, called the *corium*. Close inspection will reveal a 4-segmented beak and 3-segmented tarsi. Simple eyes, or ocelli, are absent, and the head is shorter than the prothorax. Plant bugs are less than 10mm long.

Assassin Bugs – Family Reduviidae

Deserving of their ominous name, assassin bugs are named for the cool and calculating manner in which they stalk their quarry, deliberately raise a dagger-like beak into striking position, and then stab the victim, injecting a paralyzing saliva. Enzymes in the saliva break down the prey's internal tissues into a partially digested liquid, which the assassin bug then consumes much as one might drink a beverage through a straw. When not in use, the 3-segmented beak fits snugly into a central groove between the forelegs, again calling to mind the image of an assassin as he sheaths his weapon.

The femurs of an assassin bug's forelegs are enlarged to house the powerful muscles that hold

Below Nymphs of firebugs (*Pyrrhocoris apterus*) from southern Europe sun themselves on lime twigs. Firebugs are frequently sun-worshippers, or heliotropes. These gregarious insects derive safety in numbers and in pre-adult stage have not fully developed their wings.

the struggling prey until it is subdued. The bodies of assassin bugs may be either oval or elongated, and adults range from 12mm to 36mm in length. Their heads are always elongated, and are clearly marked by a deep groove running from one eye to the other.

Ambush Bugs – Family Phymatidae

Unlike assassin bugs, ambush bugs do not stalk their victims. Rather, they rely upon their cryptic coloration to lie in wait, usually on a flower, until a visitor seeking nectar or pollen ventures too close. Though rather small themselves (8–12mm), their stout build and enormous muscular front femurs enable them to take prey much larger than themselves, including butterflies, wasps, and bumblebees. Colored with various tones of green, yellow, and/or brown, ambush bugs typically have short, club-shaped antennae and a widely flared abdomen, extending well to the sides of their wings.

Lace Bugs – Family Tingidae

The identity of lace bugs is readily apparent from their elegantly sculptured forewings and expansive thorax. Due to the aforementioned features, these small bugs (3–6mm long) appear to have a rather unusual, rectangular shape. They are strictly plant feeders, and a species will often frequent only a single host species. Sap drips from the tiny wounds made by females as they lay eggs on the undersides of leaves, often resulting in formations like miniature stalactites as it dries.

Giant Water Bugs – Family Belostomatidae

The largest true bugs, 12–65mm in length, the giant water bugs' most obvious feature, besides size, is their powerful raptorial forelegs, which they use to grasp and hold prey as they thrust sucking mouthparts into it. They swim by means of flattened hind legs, returning to the surface regularly to breathe through two short, retractable tubes at the tip of the abdomen. Most of the larger species lay their eggs on aquatic plants, but some, particularly those of the genus *Belostoma*, cement the eggs to the back of the male, who carries them about until they hatch. The prey of giant water bugs includes other insects, tadpoles, salamanders, small fish, and even snails. You can study these water-dwelling behemoths by attracting them to lights at night.

What you can do

Building a micro aquarium

If you've been adventuresome enough to try close-up photography of insects and have achieved promising results, a new challenge awaits! While many terrestrial insects can be photographed in their natural habitats, this is virtually impossible where aquatic insects are concerned. To take satisfactory photos of these subjects, you will have to build some micro aquariums, which lend themselves well to indoor photography.

Use 5cm squares of slide mounting glass, available at photographic supply stores, and silicone bathtub caulk to make a cube that is open at the top. Glue the sides to the outside edge of the bottom section, instead of to the top, to allow room to insert a vertical glass slide into the finished aquarium to maneuver subjects into position. Use distilled water instead of pond water to ensure clarity.

Before shooting, remove all air bubbles from the sides with an artist's watercolor brush. Rinse the subject thoroughly in distilled water before transferring it to your aquarium. To achieve natural-looking photographs, make a background by swirling watercolors on a piece of paper to achieve a mixture of green, tan, and brown hues, simulating aquatic vegetation and debris.

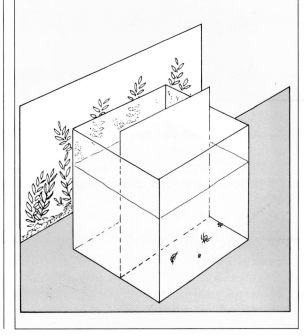

APHIDS, CICADAS, & HOPPERS
ORDER HOMOPTERA

SELECTED FAMILIES

Although entomologists once considered homopterans to be a suborder of the Order Hemiptera, the differences between the two groups have now been recognized as sufficiently significant for them to be placed in a separate order. Homoptera means "similar wing," reflecting the fact that their wings are uniformly membranous, in contrast to the unique wing design of true bugs. At rest, the wings of cicadas and hoppers slope in a roof-like manner over the body. Also different from those of true bugs are the mouthparts of homopterans, which, while again of the piercing-sucking variety and beak-like, arise far toward the posterior end of the head and can be projected down but never forward as they can among many hemipterans.

Members of this order feed exclusively on plants, but the appetites of individual species are fairly discriminating and vary tremendously. Cicadas and hoppers are vectors of some viral plant diseases, and much damage to agricultural crops and greenhouse plants is caused by the feeding of aphids and scale insects.

Cicadas and hoppers have large, razor-like ovipositors with which they cut slits and lay their eggs in the tissues of their favorite food plants. All homopterans undergo hemimetabolous development, also known as simple or incomplete metamorphosis. Nymphs resemble the adults except for their wings, which are reduced or absent.

There are about 45,000 species of homopterans worldwide, and 6,000 in North America. The larger species have distinctive features and are fairly easy to identify.

Cicadas – Family Cicadidae

Large, stocky insects, cicadas are best known for the shrill calls of the males, audible for up to a mile or more. Located on the sides of the first abdominal segment are membranous areas of the cuticle, called tymbals, which are alternately buckled and restored by the contraction and relaxation of a large muscle. Each movement of the tymbals produces a click or pulse, which in a long series constitute the cicada's call, which is a hallmark of the warm, dry days of summer. A complex series of resonating air sacs present in the abdomen greatly amplifies the call.

Cicadas are dark and fairly large – 16–60mm long – with prominently veined, membranous wings, the front pair of which are about twice as long as the hind pair. Females lay their eggs in slits made in twigs of trees and shrubs, the tips of which usually die as a result. Upon emerging, nymphs fall to the ground, where they burrow under the soil with well-developed fore legs and feed on the xylem sap of plant roots. Because of the low nutrient content of this food, development

What you can do

Cicada hunting

Excluding the years when large broods of periodic cicadas emerge simultaneously, these robust insects can be surprisingly hard to spot because of their preference for tall shade trees. Your best chance of finding them is to follow the male's raucous buzz, which swells in magnitude and then tapers off toward the end. If you cannot spy the adults, then look for their shed nymphal skins, which are hollow replicas, down to the smallest detail. You will find these still clinging to the bark of tree trunks where they were abandoned.

If you find these, there are almost certain to be burrows at the base of the tree, from which the nymphs emerged. These burrows are roughly 15mm in diameter and show no signs of excavation from the surface, such as loose dirt around the perimeter.

Above left This Australian cicada is waiting for its new wings to harden. Its body is soft also, but it will take under an hour to dry and darken.

Above This male cicada of the *Macrotristria* genus from Australia is ready to fly, since its transparent wings have hardened and its colors are fully developed.

Left Following a mass emergence of cicadas, they may all line up to dry their wings; here, a *Platypedia* species rests on a scrub oak twig in Colorado.

is slow, and the nymphs will spend the next 4 to 17 years underground, depending upon the species. Occasionally, enormous broods will emerge in the same year, leaving thousands of shed skins still clinging to the individual trees that they climbed for their final molt.

Treehoppers – Family Membracidae

Treehoppers are notable for their large pronotums that project far forward over the head, or backward over the abdomen, or both, often in bizarre protective configurations that make the insect appear as a thorn or another part of the plant upon which it feeds. They are small jumping insects, about 5mm to 12mm in length, but able to clear a meter or more in a single bound. Some are boldly marked with bright colors; in some instances, females exhibit duller colors than males, whose long-term survival is less important to reproductive success.

Treehoppers feed on the xylem sap of plants, usually trees and shrubs. Copious quantities of this nutrient-poor liquid must be ingested, and is voided as "honeydew," a sweet anal secretion relished by ants, who tend small "herds" of treehoppers much as a shepherd tends livestock. The ants get the honeydew they so desire, while the treehoppers are protected from such predators as spiders.

Spittlebugs – Family Cercopidae

Also known as froghoppers for their frog-like appearance, spittlebugs get their name from the frothy white masses manufactured by the nymphs as a moist refuge. Produced as air is blown into an anal secretion, these bubbly accumulations are commonly seen on the grasses and herbaceous plants upon which the nymphs usually feed, and sometimes on trees and shrubs as well.

Spittlebugs are small, between 4mm and 13mm in length. Grayish or brownish adults leap about freely and are easily distinguished from the similar leafhoppers by having one or two prominent spines on each hind tibia, rather than the two rows of spines that are found in the same location on leafhoppers.

Below Here disguised as plant thorns, these thornbugs (*Umbronia spinosa*), from the Peruvian rainforest, demonstrate the amazing diversity of body shape found in this species.

Below right The familiar "snake-spit" on plants is actually a protective bubbly secretion produced by the nymphs of the spittlebug, as by this member of the *Philaenus* genus.

Leafhoppers – Family Cicadellidae

Though small in size (2–15mm in length), leafhoppers constitute the largest family in the Order Homoptera. Many are brightly colored, and the two rows of spines on each hind tibia are a unique feature. Leafhoppers are found on all types of vascular plants, but are most common in meadows in other grassy areas. Most leafhoppers feed on phloem sap of plants, which is rich in organic nutrients. Among those that utilize plants' xylem sap, some are nicknamed "sharpshooters," because of the frequent and forceful method in which they expel honeydew. This family includes many plant pests; these damage the plants by their feeding and also transmit viral diseases.

Whiteflies – Family Aleyrodidae

Tiny insects, measuring only 1mm to 3mm in length, whiteflies are common outdoors in tropical and subtropical areas, but they have become pests of greenhouses and indoor plants everywhere, appearing as tiny white flecks. The wings and bodies of whiteflies are covered with a white, powdery substance. In all but the first instar, nymphs are immobile, attached to the plant by a fringe of waxy white filaments and resembling scale insects. The wings of adults are disproportionately large and fold horizontally over the body at rest.

Aphids – Family Aphididae

The sight of pear-shaped aphids, colored bright green, red, brown, or various pastel tints, is enough to strike terror into the hearts of gardeners. What makes aphids so destructive is their tremendous reproductive capacity. Unlike most other insects, which reproduce sexually, aphids can reproduce by *parthenogenesis*, a process in which the eggs develop without fertilization by a male and hatch into exact genetic replicas of the mother. In this fashion, billions of offspring can originate from just one female aphid, causing massive infestations within a short period of time. Their sheer numbers weaken and kill plants by consuming large quantities of sap and causing them to wilt and turn yellow.

A rather complex life cycle accounts for aphids' existence. Eggs that overwinter hatch into wingless females in the spring. These reproduce by parthenogenesis and are among the few truly viviparous insects, nourishing the embryos from

Top The red and black froghopper (*Cercopsis vulnerata*) is a common European species; usually found on plants, it will jump or fall to the ground if disturbed.

Above Each only a few millimeters long, gregarious whiteflies (*Aleyrodes brassicae*) are a pest in greenhouses, breeding on the undersides of leaves.

91

Above A group of aphids (*Hyalopterus pruni*), of different ages and colors, are here feeding and breeding on the stem of a reed. None of these aphids is winged, and their colors reflect their different ages and the juices they have been drinking.

Right A female nettle aphid (*Microlophium carnosum*) gives birth while drinking juices from the plant. The spines and hairs on the nettle do not appear to deter the winged adult aphids.

What you can do

Aphid ranching

It's not the Wild West, but aphid ranching will allow you to observe the mutually beneficial relationship between ants and aphids. Follow the procedure given on pages 142–3 for setting up an ant farm. Next, in late spring or summer, go hunting for a small plant bearing an aphid colony and carefully transplant this into a pot, or set several potted plants, such as roses, outdoors as "bait." (Be careful not to remove any rare plant from its natural habitat, however.) Place the entire potted plant either into an aquarium, sealing this with fine mesh and a large rubber band, or into a tall, clear plastic or cellophane bag, using a tall wooden dowel rod for support. Add the ant farm, opened to allow them access to the aphids, seal both inside, and place the ranch where the plant can get its natural amount of sunlight. Within a short time, the ants will begin to "milk" the aphids, stroking them with their antennae in a way that stimulates them to release their honeydew. When the ants have come to treat the aphids as "theirs," introduce a natural aphid predator, such as a ladybird beetle, and watch the ants' reaction.

Above Aphid-farming is a well-known aspect of the social life of ants, and here *Myrmica ruginodis* attend aphids of *Aphis viburni* on the stems of a guelder rose. It is in the interest of ants to look after aphids since, on stimulation, the aphids produce a sweet liquid on which the ants gorge themselves.

their own tissues and giving birth rather than laying eggs. After several generations, winged females appear; these migrate in a swarm to another species of plant, and continue to reproduce asexually. Late in the growing season, they return to the original host species of plant, where they produce a generation of males and females which mate and deposit another overwintering batch of eggs.

Aphids are sought out by many species of ants for the sweet anal secretions, or "honeydew," they produce. The ants zealously tend and protect them as if they were so many tiny cattle, even going so far as to carry them to the more nutritious parts of the plant.

Scale Insects – Superfamily Coccidea

This group includes several families characterized by their habit of secreting a waxy, scale-like covering on plants in order to conceal themselves or their eggs. These scaly shelters vary greatly in appearance, texture, and composition. They have been economically important, not only as plant pests, but also as the raw material from which certain red dyes, called cochineal dyes, were extracted and used to color cosmetics, medicines, and beverages. They were also the source of *lac*, from which shellac was formerly made.

Nymphs and adult females are immobile under their protective scales, while adult males have one pair of wings. A needle-like appendage at the tip of the abdomen and their lack of mouthparts sets adult male scale insects apart from gnats, which are in the Order Diptera. Identification of scale insect families is based mostly on microscopic features of the females.

What you can do

Collecting techniques	Killing methods
by hand or forceps	ethyl acetate killing jar
aspirator	immersion in 70% alcohol
sweep net	**Preservation techniques**
beating tray	pin through right side of scutellum
light trap	point
	70% alcohol vials

BEETLES
ORDER COLEOPTERA

SELECTED FAMILIES

Coleoptera means "sheath wings," a reference to the armored forewings, called *elytra*, which are either hardened or leathery and, when folded, meet in a straight line down the middle of the back, protecting the membranous hind wings that are used for flight. In flight, the elytra are held perpendicular to the body and function as airfoils, providing added lift. Many species have parallel lines or grooves called *striae* running the length of each elytron. Although nearly all beetles can fly, most do so only to cover short distances or to reach low vegetation. The rest of their time is spent either crawling on or near the ground or on vegetation or swimming.

All beetles have chewing mouthparts with well-developed mandibles that, depending on the species, are variously adapted to partake of a wide variety of foods. Most beetles are either herbivorous or scavengers, but there are also many predators among them. Many of the plant-eaters are destructive pests of crops and forests and in large numbers can cause significant damage, either directly or by transmitting diseases. Other problem beetles infest stored food, clothing, carpets, or museum specimens.

Beetles have two prominent compound eyes, and antennae of various shapes and sizes that arise between the compound eyes. The pronotum is usually quite prominent, but the other two thoracic segments, as well as most of the abdomen, are typically hidden under the elytra when viewed from the dorsal side. The metamorphosis of beetles is complete – in other words, they are holometabolous – and the larvae do not resemble the adults.

Aside from the preceding information, it is very difficult to generalize about beetles. There are more than 300,000 known species of beetles in the world, a number approximately equal to that of known plant species. They constitute about 40 per cent of all known insects, and about one-third of all known animal species. Members of this order range in size from 0.25mm to 200mm in length, up to 75mm in width, and they occur in many shapes and in all colors, although most are dark. You can find them in every conceivable habitat, except for salt water and polar ice caps. It could well be argued that beetles are the most successful organisms on earth.

Tiger Beetles – Family Cincindelidae

As their name suggests, tiger beetles are fierce predators, especially upon other insects. They are long-legged insects that run and fly swiftly, but unlike the tiger, which relies upon stealth, tiger beetles are more likely to use their speed to run down prey. Typically, they fly close to the ground for short distances.

Tiger beetles are a colorful group, sometimes adorned with an iridescent or metallic sheen, and

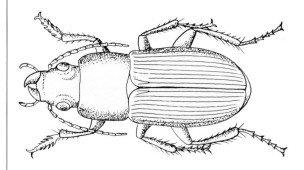

Above There are more than 300,000 known species of beetles in the world, ranging in size from 0.25mm to 200mm in length. The Latin name of their order refers to the armored forewings which meet in a straight line over their back when folded. All beetles have chewing mouthparts with well-developed mandibles, variously adapted for a wide range of food.

Above The bombardier beetle (*Brachinus crepitans*) bombards its predators with an irritating and foul-smelling aerosol, which it sprays from its abdomen when disturbed.

Left Lively tiger beetles (*Cincindela campestris*) live in sandy areas and fly off if approached too quickly. Here, a male attempts to mate; occasionally, other competing males will arrive and ride piggyback. The bright metallic colors are typical.

ranging in length from 6mm to 40mm. They are easily identified by their wide head and prominent sickle-shaped mandibles that cross in front of the head. Most are active during daylight and are fond of open, sunlit, sandy areas like beaches and paths. Handle them very carefully to avoid a painful bite.

Ground Beetles – Family Carabidae

Like the tiger beetles, with which they are sometimes lumped into a single family, most ground beetles actively pursue prey, but they are nocturnal predators. During the day, you will usually find them in moist places, under rocks, logs, or leaf litter, for example. Ground beetles are from 3mm to 36mm long, and long-legged; most are shiny and dark, although some are quite colorful.

The members of one particular group, the genus *Brachinus*, are known as bombardier beetles because of their unique defense mechanism. Two relatively benign chemicals are stored in a special abdominal chamber. When the beetle is disturbed, these chemicals are injected into another chamber, where they are acted upon by an enzyme to produce a violent reaction, releasing oxygen, water, noxious chemicals called quinones, and considerable heat. This mixture, which approaches 212°F (100°C), is released as an explosive puff of irritating gas that can be aimed by the beetle in a 360 degree radius with remarkable accuracy. The heat, irritation, and an audible report as the spray is released are enough to deter most would-be predators, and the beetle can reload and fire in rapid succession – up to a dozen times if necessary – before its reservoir is temporarily exhausted.

Predacious Diving Beetles – Family Dytiscidae

These fairly large aquatic beetles are common in most freshwater habitats. Their streamlined oval bodies are from 5mm to 70mm long and are usually dark green, yellowish-brown, or brownish-black in color. Predacious diving beetles swim by moving their hind legs in unison like oars. In this, they are unlike water scavenger beetles, with which they may be confused but which move their hind legs alternately. They must visit the surface periodically to renew their air supply, which is stored as a bubble in a special chamber under the elytra. Unlike water scavenger beetles, which break the surface film head first, predacious diving beetles come to the surface tail first.

They overwinter as adults, frequently migrating from one body of water to another in spring or fall. They fly well, and are attracted to lights at night.

Below These whirligig beetles (*Gyrinus natatior*) are about 6mm long, and are able to race over the water film in short bursts at a speed of about 40in per second.

Right The great diving beetle (*Dytiscus marginalis*) is found throughout Europe. This specimen is breathing at the water surface.

Whirligig Beetles – Family Gyrinidae

Early spring and middle to late summer are the times when you are most likely to see these shiny black or metallic dark green beetles darting in a seemingly aimless fashion across the surface film of fresh water. Whirligig beetles tend to congregate in "herds," floating serenely or gyrating in whirling patterns on calm areas of ponds, lakes, marshes, and stream eddies. They can dive underwater, and frequently do so to escape danger. The oval, flat adults are 3–15mm long, and are easy to recognize by their compound eyes, which are divided into upper and lower sections which can see both above and below the surface of the water simultaneously. Scoop-shaped segments on their short, clubbed antennae enable them to read vibrations on the surface film of the water, by which they avoid obstacles and locate prey.

Hister Beetles – Family Histeridae

These small, hard-bodied beetles are normally shiny black, but may also be bronze or green. They are less than 10mm long and their shortened elytra leave the last one or two abdominal segments exposed. Hister beetles generally live near decaying organic matter, such as carrion or dung, where they prey on other small insects. A few species live under loose bark or in termite or ant colonies.

Rove Beetles – Family Staphylinidae

The very short elytra of rove beetles, which covers only the first few abdominal segments, are most conspicuous. They are generally between 2mm and 20mm long, often with slender, ant-like bodies. Rove beetles run rapidly, with the tips of the abdomen curled up and forward like a scorpion's stinger. Despite their short elytra, they are among the best fliers of all beetles. Most are predators, and they are often found on flowers, mushrooms, under bark, or in leaf litter.

Below Rove beetles, of the Staphylinidae family, are named after their behavior of roving over the ground with the tip of the abdomen lifted upward looking for prey. One of the hind wings of this *Ontholestes tessellatus* is just visible, beneath its short elytra.

ORDER COLEOPTERA

Carrion Beetles – Family Silphidae

Insects such as carrion beetles provide one of nature's waste removal services, while at the same time enhancing the soil with their own nutrient-rich waste. Some carrion beetles excavate under the dead animal and bury it after laying eggs on it, so their offspring won't have to compete with fly larvae for food. Adults actually feed primarily on maggots on the carcass.

Carrion beetles are the largest such scavengers, ranging in length from 1.5mm to 40mm, and are usually black, with bright orange, red, or yellow markings. Somewhat flattened, with club-shaped antennae, they are easily attracted to traps baited with small dead animals. At least one species is attracted to dead animals suspended in the air, and most show a preference for certain types of animal, such as birds, snakes, mammals, fish, and so on. When threatened, some may feign death, while others emit an unpleasant odor or a foul-tasting secretion, a defense advertised by their bold markings.

Stag Beetles – Family Lucanidae

The name of this family was derived from the massive, branching mandibles on males of its larger species, which resemble the antlers of a stag and are used in similar fashion for defense and to resolve disputes over females. The beetles can use these to inflict a painful pinch if mishandled, and so are also known as pinching bugs. They range

Above The stag beetle (*Lucanus cervus*) spends up to three years as a larva eating rotting wood in old logs, and then emerges from an intermediate pupal stage into an adult about 50mm long.

What you can do

A-maze-ing beetles

If you are fortunate enough to find a large scarab beetle, especially one of the dung or carrion feeders, you can actually see them smell. Construct a small maze out of sturdy poster board or light plywood (the walls need only be centimeters high) and cover it with a clear plastic. Place a piece of dung or decaying meat as bait at one end of the maze and introduce the beetle at the other end, sealing the entrance to force the beetle into the maze. As it moves along, trying to locate the bait, you will see the plate-like processes on the tip of its antennae unfurling from tight balls as they fan out to "sniff" the air, exposing the full area of each to intercept volatile particles.

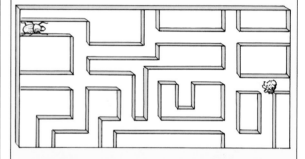

Above Carrion beetles, such as this one – *Nicrophorus vespillo*, of the Silphidae family – offer a valuable service breaking up and decaying vegetation. This specimen shows enlarged antennae, tibial spurs, and golden hairs on its body.

98

from 8mm to 60mm in length and have *lamellate* antennae, wherein the last 3 or 4 segments form a comb-like process to one side. Stag beetles are common in old woodlands, especially in decaying logs, on which the larvae feed. Despite their formidable mouthparts, adults are reported to feed on aphid honeydew and the sap of trees and leaves. Some species are attracted to lights during summer.

Scarab Beetles – Family Scarabaeidae
Scarab beetles are mostly nocturnal scavengers, playing an important role in recycling carrion, dung, and decomposing vegetation. Their front fore legs, or tibia, are wide, flattened, and sometimes toothed to serve as shovels or rakes for excavation. Members of this family are stocky, with large heads and pronotums and a highly convex body 5–60mm long; many are adorned with striking metallic colors. They have lamellate antennae tipped with comb-like processes that can

be rolled into a tight ball or unfurled. Light traps and traps baited with carrion, dung, or a fermented molasses-yeast-water mixture are effective collecting tools for scarab beetles. The Scarabaeidae includes the well-known June, Japanese, and rhinoceros beetles.

Soldier Beetles – Family Cantharidae
Soldier beetles are similar to fireflies, but they cannot produce light and lack the extended pronotum that covers the head of a firefly. The pronotum and elytra are usually red, yellow, or orange. Downy hair usually covers the elytra. Most are predatory, but a few visit flowers to feed on

Below Beetles are the most diverse group of insects on earth, and stag and rhinoceros beetles, such as this rhinoceros beetle from Mexico (*Golofa pizarro*), are among the most spectacular. The enlarged rhino spurs are for display, intimidation, and courtship fights.

pollen and nectar, and some are omnivorous. Soldier beetles are common in tall grass.

Net-winged Beetles – Family Lycidae

Members of this family have elegantly-sculptured elytra, with a delicate network of veins, interrupted by prominent, parallel ridges that run the length of each forewing. The elytra are widest toward the rear, giving the insect a pear-shaped appearance, and this effect is enhanced by the rounded pronotum, which conceals most of the head. Net-winged beetles are generally 5–19mm long, and red, yellow, or orange, often with black markings. You will usually find them during the day in densely wooded or moist areas, on flowers, foliage, or tree trunks. They are best collected by sweeping or beating the vegetation in such areas.

Checkered Beetles – Family Cleridae

A wide head with bulging eyes and a narrow, elongated body bristling with erect hairs are clues to the identity of members of the Family Cleridae. The pronotum is narrower than the elytra, and often nearly cylindrical in shape. Many sport bold markings of orange, yellow, red, or blue, sometimes in a checkered pattern. Some checkered beetles feed on pollen and are found on flowers and foliage, while many others prey upon wood-boring insects and are most commonly found on the trunks of dead or dying trees. They are rather small, usually between 3mm and 12mm in length.

Metallic Wood-Boring Beetles – Family Buprestidae

This large family is named for the metallic green, blue, copper, bronze, or black sheen of adults, which is especially noticeable on the ventral side of the body and on the dorsal side of the abdomen. The parallel sides of the body and the pointed elytra give metallic wood boring beetles somewhat of a bullet shape. Adults are attracted to dead or freshly-cut wood and tree trunks, branches, leaves, and flowers in sunlit locations, and so are especially common on the southern edge of forests in the northern hemisphere. The winding paths made in the sapwood under tree bark by their larvae, known as flathead wood borers, often kill or severely damage trees.

Click Beetles – Family Elateridae

As defenses go, that of click beetles is a surprise –

Above This metallic wood-borer, *Capnodis tenebrionis*, is now very rare in Europe. The larva spends two years underground eating the roots of plants; the adult is found in the spring on various plants.

Above and **right** The tiny elm bark beetle (*Scolytus* sp.), which is only 2mm long, carries the fungus that killed million of elms in the United States and Europe in the 1970s. Larvae from eggs laid below the bark radiate out to form galleries as the larvae become bigger. Many trees suffer from their own species of bark beetles.

quite literally. The oversized prothorax of a click beetle is loosely hinged to the rest of its body. Should a bird, reptile, or other animal seize it, the click beetle quickly arches its body at this special joint and, with a sudden flex in the opposite direction, snaps a spine-like extension of the prosternum into a snugly-fitting groove on the mesosternum. The resulting audible and tangible "click" usually startles the would-be predator into dropping the beetle which, it it lands on its back, clicks again and again, catapulting itself into the air until it lands right side up, to make its escape. Smaller click beetles are more adept at this than are their larger cousins, which may be up to 50mm long. Click beetles are easily recognized by their elongated bullet shape and the unusually large pronotum, which has pointed posterior corners.

Fireflies – Family Lampyridae

There are few spectacles in the insect world more dramatic than the sight of a meadow full of twinkling fireflies on a moonless summer night. Also known as lightning bugs, they produce their flashing luminescence in a special organ near the tip of the abdomen, where an enzyme, luciferase,

Below Wireworms, which are very common in soil and old timbers, are the larvae of click beetles. This colorful click beetle belongs to the *Alaus* genus from Kenya, and is looking for a suitable place to lay eggs. If disturbed, it immediately jumps, giving an audible click sound.

Bottom Sprung for action, this click beetle (*Chalcolepidius porcatus*) holds the extension from its pronotum into the notch from which it can be released to right itself when upside down or to startle would-be predators.

acts upon another substance, luciferin; not all species, however, are capable of producing light. The color, frequency, and sequence of the flashes are unique to individual species, and help males and females to locate and recognize one another. Most of the fireflies that we see are males, which fly around open areas while emitting their specific signal. Upon recognizing the correct code, a female perched on or near the ground will respond, and the male quickly homes in on her beacon. The females of a few species mimic the signals of other species and devour the males.

Although there are about 2,000 species of this family in the world, fireflies are best represented in the Americas and Asia, the best example in Western Europe is the glowworm. Fireflies are medium-sized insects, from 5mm to 20mm in length, with rather soft elytra and a forward extension of the pronotum that shields the head. Females may be short-winged or wingless. Those that emit light are obviously nocturnal, and each species flashes actively only during a limited period of the night. Much of the rest of their time is spent clinging to foliage, tree trunks, branches, or in moist places under bark or decaying vegetation.

What you can do

Collecting tips
Capture techniques
by hand or forceps
aspirator
sweep net
aerial net
aquatic net
beating tray
bait trap
light trap
Berlese funnel

Killing methods
ethyl acetate killing jar
immersion in 70% alcohol

Preservation techniques
pin through right elytron
point
vial of 70% alcohol

Note Certain large members of the ground beetle and scarab beetle family develop greasy deposits on their surface after pinning. Remove the labels from the pin and degrease the beetle by immersing it repeatedly in a small amount of organic solvent, such as alcohol or acetone, replacing the solvent until it no longer becomes discolored.

Ladybird Beetles – Family Coccinellidae

Oft portrayed in children's books and nursery rhymes, the amicable ladybird beetle is one of our greatest insect allies. Also known as ladybugs and lady beetles, these diminutive – between 1mm and 10mm long – dome-shaped insects and their larvae are voracious predators upon such insect pests as aphids, scale insects, mealybugs, and mites. So effective are they that many garden supply companies sell them as biological control agents. Their round, convex, and colorful forms and their abundance make them one of our most familiar insects, despite their small stature. Most are red, orange, or yellow with black spots, but this color pattern is reversed in some species, and others are spotless. They frequently overwinter as adults in huge swarms under bark and leaf litter, a practice that seems to produce and conserve warmth.

Above Glowworms (*Lampyris noctiluca*) belong to the same sub-family of beetles as fireflies, and have a wide distribution in Europe. The light is produced only by females, as a means of attracting the winged males.

Above Ladybirds of the 11-spot species (*Coccinella 11-punctata*) gather together during their late summer sleep and during hibernation over winter. They often choose to squeeze into tree stumps and cracks in wood, where they remain dormant for several months. They may seek their winter refuges at the tops of mountains, and can be found gathered together under stones.

103

Darkling Beetles – Family Tenebrionidae

A heavily striated or bumpy elytra is the hallmark of these mostly dull black or brown beetles, and close inspection will reveal their compound eyes to be kidney-shaped or notched, rather than round. Most are nocturnal scavengers, found under rocks, debris, or decaying wood, but some species infest and damage stored food, clothing, rugs, and museum specimens. They can also damage insect collections, a good reason to keep yours in a sealed case! Darkling beetles are slow-moving, small to medium in size (2–35mm long), with variable body shapes.

Long-horned Beetles – Family Cerambycidae

Long-horned beetles are named for their long, swept-back antennae, which, combined with their often brilliant colors, make them popular with collectors. The vast majority of species in this family have antennae that are at least half as long as their elongated, cylindrical bodies, and may be up to three times the body length, which itself measures 6–75mm. The larvae tunnel through the wood of dead, dying, and occasionally living trees to feed, then bore into the bark to pupate. Most adults feed either on the pollen, nectar, and stamens of flowers, or on foliage, fruit, sap, roots, or fungi.

Though they are the bane of the timber industry because of the defects their larvae cause in recently-felled trees, long-horned beetles accelerate the decomposition of dead trees, and are therefore a vital element in forest ecosystems.

Leaf Beetles – Family Chrysomelidae

True to their name, leaf beetles are herbivorous. Adults feed on flowers and foliage, but most larvae eat leaves or bore into roots and stems. The larvae of some species are leaf miners, tunneling between the outer layers of leaves and feeding on the spongy tissue, which often results in leaf deformities. They are more common in weedy, open areas and on bushes than on trees.

Though common and often brightly colored, leaf beetles can be somewhat difficult to identify because of their varying body shapes. Some resemble long-horned beetles, though their antennae are nearly always less than half as long as their body. Others look much like ladybird beetles, and these can be much more difficult to separate, sometimes requiring scrutiny of the tarsal

What you can do

Observing metamorphosis

One species of darkling beetle, *Tenebrio molitor*, is a particularly good specimen with which to study the stages of complete metamorphosis, also known as holometabolous development. Their larvae, known as yellow mealworms, are sold in pet stores as food for fish, reptiles, and amphibians, and so are readily available. An old goldfish bowl or large wide-mouthed jar makes an excellent home for breeding mealworms. Place a layer of bran or oatmeal, about 1cm deep, in the bottom of the jar, followed by an equal layer of dried bread or cookie crumbs mixed with flour. Top this with another layer of bran or oatmeal, and introduce several dozen mealworms. They will need a small amount of moisture, which can be supplied with a slice of apple or carrot, replaced daily. Add a piece of crumpled paper towel or tissue for the nocturnal adults to hide under, and to ensure adequate ventilation secure a muslin or similar cover over the top of the jar with a rubber band. With this simple set-up, you can raise many generations of the beetles and make detailed observations of their life history.

crumpled paper towel

slice of apple or carrot

mealworms

bran or oatmeal

dried bread or cookie crumbs mixed with flour

bran or oatmeal

segments, which number three in ladybird beetles and four in leaf beetles.

Weevils – Family Curculionidae

Most weevils have a prominent, down-turned snout, protruding forward from the head, at the end of which chewing mouthparts are located. Most species have elbowed, clubbed antennae, extending laterally from midway on the snout, which often has a groove to accommodate the long first antennae segments. Ranging in length from 1mm to 40mm, weevils are the largest family in the animal kingdom, with more than 40,000 species worldwide. All are plant eaters, and many are significant agricultural pests.

Right Long-horned beetles (family Cerambycidae) are named after their antennae, which are often longer than their bodies; this one, *Penthea pardalis*, is from a eucalyptus forest in Queensland, Australia.
Below left The remarkable protrusion of this weevil (*Curculio glandium*) is typical of this group of beetles (also known as snout beetles). The sensory pair of antennae are clearly seen.

Below right The Colorado potato beetle (*Leptinotarsa decemlineata*) is a native of the Americas, though it has now been introduced to Europe where it continues to be a pest on potatoes. Other varieties are yellow and black.

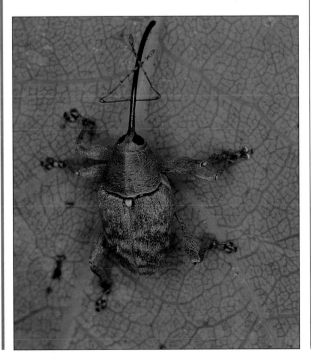

KEY TO COMMON BEETLE FAMILIES

Each step in the key asks a question about the specimen, usually answered by "yes" or "no," which then directs you to another part of the key or reveals the family name.

1 Does the front of the head extend into a long and narrow or broad and short snout, or is the body cylindrical and the antennae elbowed and club-shaped?
a. Yes .... Go to #74
b. No .... Go to #2

2 Is the insect aquatic, with hind coxae flattened into large plates?
a. Yes .... Haliplidae
b. No .... Go to #3

3 Is the posterior edge of the first abdominal sternite interrupted by the hind coxae?
a. Yes .... Go to #4
b. No .... Go to #8

4 Are the hind legs flattened and hairy?
a. Yes .... Go to #5
b. No .... Go to #6

5 Are the antennae long?
a. Yes .... Gyrinidae
b. No .... Dytiscidae

6 Is there a groove crossing the metasternum in front of the hind legs?
a. Yes .... Go to #7
b. No .... Rhysodidae

7 Is the head, including the compound eyes, as wide or wider than the pronotum?
a. Yes .... Cincindelidae
b. No .... Carabidae

8 Are the tarsi of the front and middle legs 5-segmented and the hind tarsi 4-segmented?
a. Yes .... Go to #49
b. No .... Go to #9

9 Are all tarsi 5-segmented?
a. Yes .... Go to #11
b. No .... Go to #10

10 Are all tarsi 3-segmented?
a. Yes .... Go to #72
b. No .... Go to #62

11 Do the antennae have clubbed, comblike projections?
a. Yes .... Go to #12
b. No .... Go to #18

12 Do the antennae have many comblike projections?
a. Yes .... Go to #13
b. No .... Go to #14

13 Is the ventral side hairy?
a. Yes .... Scarabaeidae
b. No .... Rhipiceridae

14 Can the last 3 or 4 plates of the clublike antennae be closed into a ball?
a. Yes .... Scarabaeidae
b. No .... Go to #15

15 Is the last tarsal segment much longer than the others?
a. Yes .... Go to #16
b. No .... Go to #17

16 Is the insect aquatic?
a. Yes .... Dryopidae
b. No .... Silphidae

17 Are the elytra marked by conspicuous longitudinal grooves?
a. Yes .... Passalidae
b. No .... Lucanidae

18 Are the hind legs curved and flattened for swimming?
a. Yes .... Hydrophilidae
b. No .... Go to #19

19 Do the elytra cover less than half the length of the abdomen?
a. Yes .... Staphylinidae
b. No .... Go to #20

20 Do the elytra cover all abdominal segments?
a. Yes .... Go to #21
b. No .... Scaphidiidae

21 Are fewer than seven ventral segments of the abdomen visible?
a. Yes .... Go to #24
b. No .... Go to #22

22 Do the middle coxae touch or almost touch each other?
a. Yes .... Go to #23
b. No .... Lycidae

23 Is the head concealed under the pronotum?
a. Yes .... Lampyridae
b. No .... Cantharidae

24 Are six ventral segments of the abdomen visible?
a. Yes .... Go to #25
b. No .... Go to #27

25 Is the body flattened and oval?
a. Yes .... Silphidae
b. No .... Go to #26

26 Are the elytra wedge-shaped?
a. Yes .... Melyridae
b. No .... Cleridae

27 Are the antennae both club-shaped and bent in an elbow?
a. Yes .... Histeridae
b. No .... Go to #28

28 Is the body dark orange or light brown and densely covered with yellow, gray, or white hair?
a. Yes .... Byturidae
b. No .... Go to #29

29 Is more than half of the head concealed under the pronotum when viewed dorsally?
a. Yes .... Go to #30
b. No .... Go to #37

30 Are the antennae at least half as long as the body and filiform (threadlike) in shape?
a. Yes .... Go to #31
b. No .... Go to #32

31 Do any of the antenna segments have projections at the base?
a. Yes .... Ptilodactylidae
b. No .... Ptinidae

32 Is the body elongated, narrowly oval, or cylindrical?
a. Yes .... Go to #33
b. No .... Go to #36

33 Are most antennae segments broader than they are long?
a. Yes .... Dryopidae
b. No .... Go to #34

continued

Above Beetles, with over 300,000 species, are the most successful order of organisms to inhabit the earth. One reason for this is their varied methods of defense. These blister beetles (*Tetraonyx frontalis*) from Mexico utilize bright colors to advertise the toxic chemical cantharidin in their tissues, which can cause blistering of the skin.

Continued from p107

34 Are the pronotum and elytra wrinkled in appearance, with curved ridges alternating with rows of puncture-like marks?
a. Yes .... Elmidae
b. No .... Go to #35

35 Is the pronotum armed with knobs, hooks, or teeth near the front?
a. Yes .... Bostrichidae
b. No .... Anobiidae

36 Are the last three or more antennae segments longer than the others?
a. Yes .... Anobiidae
b. No .... Helodidae

37 Do each of the tarsi have one or more lobes on the ventral surface?
a. Yes .... Go to #38
b. No .... Go to #41

38 Is the fourth tarsal segment the only one with a ventral lobe?
a. Yes .... Helodidae
b. No .... Go to #39

39 Do the first, second, third, and fourth tarsal segments all have ventral lobes?
a. Yes .... Cleridae
b. No .... Go to #40

40 Are the posterior corners of the pronotum distinctly pointed?
a. Yes .... Elateridae
b. No .... Dascillidae

41 Are the front coxae cone-shaped and prominent?
a. Yes .... Go to #42
b. No .... Go to #43

42 Is the beetle aquatic, with serrate antennae?
a. Yes .... Psephenidae
b. No .... Dermestidae

43 Are the front coxae at right angles to the anterior-posterior axis of the body?
a. Yes .... Go to #44
b. No .... Go to #45

44 Is there a gap between the posterior edge of the pronotum and the base of the elytra?
a. Yes .... Trogositidae
b. No .... Nitidulidae

45 Does a spine extend between the front coxae from the posterior edge of the prosternum?
a. Yes .... Buprestidae
b. No .... Go to #46

46 Are the last two or three segments of each antenna significantly larger than the others, and does the head narrow behind the eyes?
a. Yes .... Lycidae
b. No .... Go to #47

47 Are the cavities from which the coxae protrude touched by the lateral sclerites of the mesothorax?
a. Yes .... Cucujidae
b. No .... Go to #48

48 Is the pronotum much wider than the head, with the anterior corners somewhat pointed?
a. Yes .... Erotylidae
b. No .... Languriidae

49 Are the cavities from which the front coxae protrude closed behind the lateral sclerites of the pronotum?
a. Yes .... Go to #50
b. No .... Go to #52

50 Is the body covered with short, fine hair?
a. Yes .... Allecullidae
b. No .... Go to #51

51 Is the last segment of the antennae about the same size as the other segments?
a. Yes .... Tenebrionidae
b. No .... Lagriidae

52 Are the lateral edges of the pronotum rounded?
a. Yes .... Go to #56
b. No .... Go to #53

53 Is the body long and very flat?
a. Yes .... Cucujidae
b. No .... Go to #54

54 Is the body higher than it is wide, with the head tilted downward to produce a hunch-backed appearance?
a. Yes .... Mordellidae
b. No .... Go to #55

55 Are there two depressions located near the posterior edge of the pronotum?
a. Yes .... Melandryidae
b. No .... Cryptophagidae

56 Are there two depressions located near the posterior edge of the pronotum?
a. Yes .... Melandryidae
b. No .... Go to #57

57 Is the pronotum widest in front and narrower than the elytra in back?
a. Yes .... Oedemeridae
b. No .... Go to #58

58 Are at least the terminal portions of the antennae filiform (threadlike) in shape?
a. Yes .... Go to #59
b. No .... Pyrochroidae

59 Is the pronotum widest at the middle and narrow in the front and back?
a. Yes .... Salpingidae
b. No .... Go to #60

60 Does the abdomen show six ventral segments?
a. Yes .... Meloidae
b. No .... Go to #61

61 Do the hind coxae touch each other?
a. Yes .... Pedilidae
b. No .... Anthicidae

62 Is there a row of conspicuous, flat spines on the outer edge of each tibia?
a. Yes .... Heteroceridae
b. No .... Go to #63

63 Are the antennae distinctly club-shaped?
a. Yes .... Go to #64
b. No .... Go to #69

64 Are the antennae elbowed?
a. Yes .... Scolytidae
b. No .... Go to #65

65 Is the body dark, shiny, oval, and very convex?
a. Yes .... Phalacridae
b. No .... Go to #66

66 Is the head concealed by the pronotum when viewed from above?
a. Yes .... Ciidae
b. No .... Go to #67

67 Are the antennae clavate (segments gradually enlarged toward the tip)?
a. Yes .... Mycetophagidae
b. No .... Go to #68

68 Are the anterior corners of the pronotum curved prominently forward?
a. Yes .... Endomychidae
b. No .... Erotylidae

69 Is the body oval and highly convex?
a. Yes .... Byrrhidae
b. No .... Go to #70

70 Are the antennae longer than half the body length?
a. Yes .... Cerambycidae
b. No .... Go to #71

71 Is the body egg-shaped and widest toward the rear?
a. Yes .... Bruchidae
b. No .... Chrysomelidae

72 Are the elytra short, concealing only about half of the abdomen?
a. Yes ....Pselaphidae
b. No .... Go to #73

73 Is the body shape broadly oval to almost spherical and highly convex dorsally?
a. Yes .... Coccinellidae
b. No .... Endomychidae

74 Does the front of the head extend into a long and narrow or short and broad snout?
a. Yes .... Go to #75
b. No ....Scolytidae

75 Is the snout short and broad?
a. Yes ....Anthribidae
b. No .... Go to #76

76 Is the snout curved downward and are the antenna club-shaped and elbowed?
a. Yes .... Curculionidae
b. No .... Brentidae

Above True to their name, water scavenger beetles, like *Hydrophilus piceus* from England, are mostly aquatic and feed on decaying plants. Unlike predacious diving beetles, with which they may be confused, water scavenger beetles kick their hind legs alternately while swimming and surface head first for air.

BUTTERFLIES & MOTHS
ORDER LEPIDOPTERA

SELECTED FAMILIES

What could enhance the serenity of a sunny meadow full of wildflowers more than the occasional butterfly flitting aimlessly about? A kaleidoscope of colors and patterns, delicate structure, and ethereal flight combine to elevate butterflies above all other insects in the eyes of most people. They are truly the crown jewels of the insect world. Moths, too, enjoy a somewhat heightened status, even though their image is a bit tarnished by the irritating habits of a few species and many are nocturnal and therefore less familiar than butterflies.

Members of the Family Lepidoptera, which means "scale wing" when translated from its Greek roots, are characterized by two pairs of large, membranous wings covered in roof-shingle fashion by tiny scales that account for their often brilliant hues. These colors are the result of either pigments in the scales, or their prismatic structure, or both. The configuration of scales may be such that they refract, or bend, white light to produce blue, iridescent, or metallic colors. Pigments absorb all wavelengths of visible light except for one which is reflected and seen as a given color. Wing scales of butterflies and moths rub off easily, so all specimens should be handled very gently.

At rest, most butterflies hold their wings vertically over the body, while moths hold theirs either in a roof-like slope over the body, or curled around it, or spread flat against the support to which the moth is clinging. All butterflies are active only during daylight, and most moths are nocturnal, although the more colorful fly by day.

The adults of most species have siphoning mouthparts that, at rest, are coiled like a watch spring under the head. These mouthparts can be unfurled into a long tube, or *proboscis*, capable of probing deep into flowers in search of nectar, and certain species also use their proboscis to feed on tree sap or the juices of fermenting fruit. Butterflies and moths are important pollinators of many plants. In some species, the mouthparts of adults are underdeveloped or absent, so these individuals do not eat, and the adults of a few of the more primitive moths use chewing mouthparts to consume pollen. Butterfly antennae are always long, slender, and usually knobbed or hooked at the tip, while those of moths vary but are often feathery or fan-like.

Both moths and butterflies are holometabolous. The wormlike larvae, commonly known as *caterpillars*, have strong chewing mouthparts. The vast majority are plant-eaters, and the habits of a few

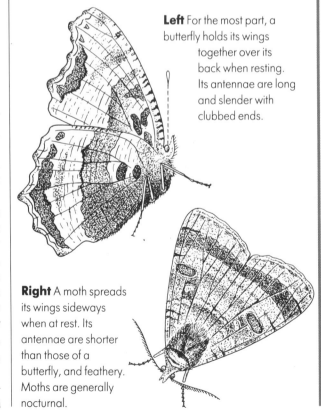

Left For the most part, a butterfly holds its wings together over its back when resting. Its antennae are long and slender with clubbed ends.

Right A moth spreads its wings sideways when at rest. Its antennae are shorter than those of a butterfly, and feathery. Moths are generally nocturnal.

Above With wings held close together over its body, this danaid butterfly (*Danaus eresimus*) from Mexico exhibits a typical butterfly characteristic. Also characteristic are the clubbed tips of the antennae.

Left The wings of this hawk moth (*Batocnema coquereli*), from Madagascar, are held flat across the body exhibiting one of the moth characteristics. Moths are generally drab-colored compared to butterflies. The colors of this rather bright moth break up its outline when on tree trunks.

species have earned them the status of "pest." Caterpillars of most species are restricted to only one species of food plant or to a closely related group of plants.

Caterpillars have a pair of true legs on each of the first three segments immediately behind the head, which together constitute the thorax. Farther back, on the abdominal segments, they also have as many as five pairs of pseudo-legs, or *prolegs*, which are reabsorbed during pupation. Some species are arrayed with stinging hairs or horn-like structures for defense.

The butterfly pupa, which is called a *chrysalis*, is sheathed in a hardened shell that is attached to or near the larvae's food plant. In contrast, the moth pupa is either encased in a cocoon of silk woven by the caterpillar and attached to a support, sheltered in a chamber within a plant stem or underground, or buried beneath leaf litter.

Upon emerging from the chrysalis, most adults set about their primary task, that of reproduction. Lepidopterans are unusual among insects in that males and females of many species exhibit *sexual dimorphism*, meaning a difference in the colors and/or patterns of males and females, indicating that discrimination between the sexes in these species is largely visual. Among those not so endowed, scent disseminated from glands on the

wings and/or abdomen is the primary factor in determining sex.

Color also serves as a means of defense for these large-winged insects. Cryptic species often fly at night and are camouflaged against their typical background by day, while those that are brightly colored are advertising the distasteful or toxic nature of their tissues to predators. Many families in this order display "eyespots" on their wings; these really do bear a remarkable resemblance to eyes. It is thought that they also provide a means of defense, because the eyespots may divert a predator's attack toward what appears to be the head but is in fact away from the insect's body. Eyespots can also be flashed suddenly from concealment to startle and frighten small predators.

The span of adulthood in butterflies and moths ranges from about one week to approximately eight months, but the average period for most species is reported to be two to three weeks, during which time courtship, mating, and egg-laying must occur. Courtship can be an elaborate affair, sometimes involving ritualized flights and/or wing stroking between potential mates. Copulation itself is sometimes achieved in flight and may last for several hours.

Eggs, which vary greatly in shape, color, and texture, are normally laid on the preferred food plant of the larvae. In temperate regions, eggs of the last generation of the year either hatch in the autumn, in which case the caterpillars must find a sheltered niche and spend winter in a dormant state in their larval or pupal form, or they hatch during the following spring. A few species over-winter as adults, finding shelter in autumn and entering a dormant period during the colder months.

A small number of species migrate to escape the rigors of winter. One notable example, the monarch butterfly, *Danaus plexippus*, migrates many hundreds of miles to its wintering grounds in southern North America, with the following generations returning to northern breeding grounds the next spring.

Skippers – Family Hesperiidae
Technically neither butterflies nor moths, skippers have some traits of each group. The body of a skipper is thick and bulky, as opposed to that of a true butterfly, which has a long and slender body

---

What you can do

Collecting tips
Capture techniques
aerial net
bait
light trap
Killing method
ethyl acetate killing jar
Preservation techniques
pin adults, wings spread, through right side of notum

Special notes
1 Do *not* kill members of this family unless you are a serious collector.

Butterflies and moths are one of the few groups of insects whose populations have suffered major declines due to human activities.

2 Reserve a special killing jar for Lepidopterans, so that other specimens do not get contaminated by the lost wing scales.

3 Place each specimen in a triangular paper envelope (see page 55) for protection until they can be pinned.

Above The alert little silver-spotted skipper (*Hesperis comma*) ventures from the woodland edge to a garden zinnia to suck up nectar. Its hooked antennae are characteristic of skippers, and the silver spots on the greenish-tinged underside of the hind wing are typical of the species.

Right The European "scarce swallowtail" (*Iphiclides podalirius*) is not particularly scarce in southern Europe, compared to the ordinary swallowtail, *Papilio machaon*. Its tails and eye-spot mimic the head, compound eye and antennae, thus confusing predators.

in comparison. Skippers are characterized by a wide head and hooked antennae that are widely separated at the base. They can be recognized at a distance by their habit of resting with forewing spread at a 45-degree angle and hind wings spread 180 degrees in a horizontal plane. Wingspans in this family generally range between 14mm and 50mm. Their name can undoubtedly be attributed to their swift, direct, and "skipping" flight.

Swallowtails – Family Papilionidae

Swallowtails are probably the most familiar butterfly family, and possibly the best known family of insects, in the world. They are large, even by butterfly standards, with wingspans of between 54mm and 150mm. Most are either yellow or white, with bold black markings and bright yellow, orange, red, or blue spots, or are black with similar spots. This group is named for the rearward extensions on the hind wings of some species,

which resemble the forked tails of swallows. Parnassians, which lack the tail-like projections of swallowtails, can still be recognized as members of this family by their hind wings, which have only one anal vein.

Milkweed Butterflies – Family Danaidae

Although considered by some to be part of the Family Nymphalidae, milkweed butterflies have antennae completely without scales and therefore, in the opinion of entomologists, differ significantly from them. Milkweed butterflies are fairly large, with wingspans of between 75mm and 102mm. Males have scent pouches on the hind wings and, in some species, brush-like tufts within the abdomen that can be extended to release pheromones that have a tranquilizing effect on females during mating. Most species in this family occur in Asia. North American species are orange or brownish-orange, with black markings; the most famous of

Right The bold orange colors of the North American monarch (*Danaus plexippus*) are a warning to predatory birds that the taste is not to their liking.

Below The clouded yellow (*Colias croceus*) is a very common migratory butterfly. Its pink legs and antennae are a means of camouflaging this butterfly from spiders that lurk in pink-colored flowers, such as thistles.

these is the amazing migratory monarch.

Milkweed butterflies take their name from the larval food plants of North American species. Toxins in the milky sap of these plants are incorporated into the tissues of the insect, imparting a disagreeable taste and causing uncontrollable retching in birds and other predators that ingest them. A number of species outside this family (the Viceroy, for example) mimic the colors and patterns of milkweed butterflies, gaining passive protection for this reason.

Satyrs, Nymphs, & Arctics – Family Satyridae

The wings of satyrs and their kin span between 25mm and 73mm, and are mostly drab gray or brown, with brightly colored eyespots or other markings. Like their close relatives, the brush-footed butterflies, members of this family have front legs that are greatly reduced in size and are not used for walking. The chief distinctions between these two families is that satyrs have from one to three greatly enlarged veins on the front wings. Their erratic, dancing flight takes them

What you can do

Butterfly photography

Butterflies are among nature's most photogenic subjects. It is more difficult to capture these elusive creatures on film than to collect them, but a successful photograph makes a beautiful portrait of the insect in its natural habitat, and it does the butterfly absolutely no harm, an important consideration with less common species. Also, you can photograph all stages of a butterfly's life cycle and its food plants much more easily than you could represent them in a collection.

Successful butterfly photography demands the use of a macro lens, which allows you to focus close enough to obtain frame-filling shots of small subjects. A macro lens should yield at least 1:1 reproduction. That is to say, the image on the film

will be life-size. The major difference between macro lenses is their focal length. A 55mm macro lens can yield about the same image as you will achieve by using a standard 50mm lens to shoot subjects, such as landscapes or people; the difference is that you must be very close to shoot macro subjects. A 105mm macro lens will give you a greater working distance between the camera and the subject, and a 200mm macro allows even more – no small advantage when working with such wary subjects as butterflies.

Using an electronic flash may help to fill in details and colors, but try to use an exposure as close as possible to that dictated by natural light, so that your background consists of flowers and foliage rather than blackness.

Above This photograph of a *Pieris rapae* in flight shows what can be achieved with a 50mm lens and one extension. High speed films enable you to freeze movement and some of the new automatic cameras will give you a shutter speed of 1/2000 of a second.

---

weaving close to the ground through the vegetation of their wooded or open, brushy habitats. All caterpillars feed on grasses and/or sedges.

Whites, Sulphurs, & Orange Tips – Family Pieridae

This family runs a close second to the swallowtails in terms of the general public's familiarity with them; in fact, the name "butterfly" was probably derived from the butter-yellow color of familiar European sulphurs. Distribution is worldwide, and they are found over open, sunny areas such as meadows and fields. The names are descriptive of their typical white, yellow, or orange wing colors, which usually bear either simple black markings or none at all. Orange tips are relatively uncommon and have marbled greenish-yellow, orange tipped wings. With wingspans of between 22mm and 70mm, this is a family of small to medium-sized butterflies. The tips of their otherwise slender antennae are abruptly swollen. Individuals of the same species often exhibit variations in color, and the sexes of many species exhibit sexual dimorphism, a difference in pattern and color between males and females. Some species, such as the Cabbage White, consume food crops.

Gossamer-winged Butterflies – Family Lycaenidae

This family includes those butterflies commonly known as harvesters, coppers, blues, and hairstreaks. Most have dazzling blue, green, violet, or coppery colors, all of which are due to light-bending microscopic structures on the scales. Although similar to metalmarks of the Family Riodinidae, gossamer-winged butterflies hold their wings vertically over the back when at rest, whereas metalmarks usually keep theirs spread flat or at a 45-degree angle. The former are fairly small as butterflies go, with wingspans of from 11mm to 51mm and a brisk, darting flight. Males have front legs that are reduced in size.

Below The colors and patterns of butterflies are fascinating; here, the male of the common blue (*Polyommatus icarus*) shows off its blue iridescence, which is a structural color caused by light refractions.

Right The zigzag patterns, false eye, and two sets of tails on the dusky blue hairstreak (*Calycopis isobeon*), from Mexico, contribute to its head-to-tail mimicry.

Bottom The astoundingly lovely purple-edged copper (*Palaeochrysophanus hippothoe*) from southern Europe takes on a wonderful iridescence in strong sunlight. This specimen was photographed at 6,000 feet in the French Pyrenees.

What you can do

Creating a butterfly garden

Since the middle of the 20th century, the increasing and indiscriminate use of pesticides, combined with accelerated habitat loss, have dealt butterfly populations two heavy blows. By planting a butterfly garden, you can help them while enjoying both their beauty and that of wildflowers. Depending upon your ambition and resources, and the space that you can make available, your butterfly garden can range in size from a few window boxes or a patio border up to a full-blown meadow.

Your first step should be to take a survey of public gardens, parks, meadows, vacant lots, and similar sunny places around your home during spring, summer, and fall and, with the aid of field guides, identify the plants that are already attracting butterflies to the area. These should form the basis of your garden, and to them you can add food plants listed in the butterfly guides for those species normally found in your region that you would like to attract but have not observed. To maximize the effective season of your butterfly garden, select plants with overlapping blooming periods.

You can either buy seeds or plants commercially or gather seeds from wild plants, remembering to observe all local regulations. Seeds gathered in fall should be planted immediately or stored outdoors in a dry place that simulates natural winter conditions as closely as possible. Consult reference books for information on growing native plants from seed. Natural history museums, botanical gardens, and university botany and entomology departments are also good sources of information.

The following spring is the time to plant your garden. Most butterflies prefer sunny, open locations, so keep this in mind when selecting a site. Individual species will seek nectar at different times of the day, so scatter the various plants throughout the garden to ensure that at least some will be sunlit when a given species is feeding. Keeping taller plants toward the perimeter will also help.

In addition to nectar sources for adults, a complete butterfly garden will also include food plants for caterpillars, mud and water puddles, rocky areas, and shelter from rain and wind.

Some popular butterfly food plants

Common Name	Genus	Adult	Cater-pillar
alfalfa *	Medicago spp.	X	
apple	Malus spp., Pyrus spp.		X
aspen	Populus spp.		X
aster *	Aster spp.	X	
bee balm	Monarda didyma	X	
beggar ticks	Bidens spp.	X	
blazing star	Liatris spp.	X	
buttonbush	Cephalanthus spp.	X	
butterfly bush *	Buddleia spp.	X	
cherry	Prunus spp.		X
clover *	Trifolium spp.	X	X
columbine	Aquilegia spp.	X	
daisy	Chrysanthemum spp.	X	
dandelion	Taraxacum spp.	X	
dill	Foeniculum vulgare		X
dogbane *	Apocynum spp.	X	
fennel	Anethum graveolens		X
goldenrod	Solidago spp.	X	
grasses	Poa spp.		X
hackberry	Celtis spp.		X
hops	Humulus spp.		X
ironweed	Vernonia spp.	X	X
jewelweed	Impatiens spp.	X	
knapweed	Centaurea spp.	X	X
lantana	Lantana spp.		X
lilac	Syringa spp.	X	
mallow	Malva spp.	X	X
milkweed *	Asclepias spp.	X	X
mint *	Mentha spp.	X	
nettle	Urtica spp.		X
parsley	Petroselinum crispum		X
penstemon	Penstemon spp.	X	
plum	Prunus spp.		X
poplar	Populus spp.		X
privet	Ligustrum spp.	X	
rock cress	Arabis spp.		X
self-heal	Prunella vulgaris	X	
sweet pepperbush	Clethra alnifolia	X	
thistle *	Cirsium spp.	X	X
tickseed	Coreopsis grandiflora	X	
vetch	Vicia spp.		X
violet	Viola spp.	X	X
wild carrot	Daucus carota	X	X
willow	Salix spp.		X
winter cress	Barbarea spp.	X	X
wormwood	Artemisia spp.	X	
yarrow	Achillea spp.	X	

* extremely popular with adult butterflies

What you can do

Butterfly watching tools

aerial net
field guides (to butterflies, wildflowers, and trees)
glass or plastic jar
hand lens
blunt forceps
binoculars
field notebook

For the environmentally-minded practical entomologist, butterfly watching is an attractive alternative to collecting.

Note the flight patterns of individuals, as this will often be an important clue to their identity. You may need to capture the butterfly to make a positive identification. With practice, this can be done without harming it. Upon netting a butterfly, give the net handle a quick half twist to trap the insect inside and gently immobilize it as quickly as possible. You may remove your captive from the net by grasping the closed wings with a pair of blunt forceps (never with your bare fingers, as too many scales will be rubbed off) and place it in a jar for observation with a hand lens as you compare it with descriptions and illustrations in the field guide. You will probably want to observe behavior before capturing a specimen, since it may be some time before a released butterfly returns to its normal activities.

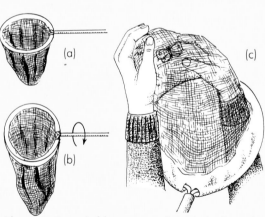

Above At the end of the sweep twist your wrist (a) to fold the bag over the mouth of the net (b). A carrying container can be made by taping a thick piece of black paper over one end of a cardboard tube. Insert this into the net (c), perch the butterfly on your finger and maneuver it carefully into the container. Seal with black cloth and an elastic band.

Metalmarks – Family Riodinidae

Although the majority of this family are colorful species found in the tropics of the western hemisphere, the wings of most temperate region metalmarks are colored in shades of brown, gray, or rust. Many species in the Family Riodinidae also display shiny metallic lines, borders, or other marking. They are known for their habit of perching on the undersides of leaves, with their wings spread flat or at a 45-degree angle. Like gossamer-winged butterflies, males of this family also have vestigial front legs that are not used for walking. Metalmarks are small butterflies, with wingspans of from 16mm to 51mm.

Brush-footed Butterflies – Family Nymphalidae

A very diverse family, brush-footed butterflies have been located in nearly all terrestrial habitats except polar ice caps. They are medium-sized somewhat orange butterflies with wingspans usually ranging between 36mm and 78mm. Most are good fliers, and several migrate or otherwise overwinter as adults. The common name of this family is apparently a reference to their vestigial pair of front legs, which are too short to be used for walking. Brush-footed butterflies are also distinguished by conspicuous knobs on their antennae and by their thick, hairy palpi. The family includes painted ladies, tortoiseshells, admirals, checkerspots, fritillaries, longwings, and crescentspots.

Above Snout butterflies are named after the extension of their mouthparts. There are many species in the Americas, and this one (*Libythea carinenta*) from Mexico is feeding on wild fruit. Typically, snout butterflies hibernate through the winter and emerge in the spring to suck juices from the buds of trees and shrubs.

Right The painted lady (*Cynthia cardui*) is one of the world's most powerful butterflies. During the spring and summer it moves northward in waves from North Africa, through Western Europe. Once it finds a suitable place, the butterfly often defends its own territory. This female is swollen with eggs, which she will lay on thistles.

Above Dusted with iridescent scales, this metalmark (*Caria mantinea*, of the Nemobiidae family), from Peru, displays the typical characteristics of its family. It rests with its wings held flat, rather like a moth.

119

Sphinx Moths – Family Sphingidae

Sphinx moths, also known as hawk moths, constitute a large and fairly common family. They are stout-bodied and have rigid, narrow, and often boldly marked wings. The front wings are normally about twice as long as the hind pair, spanning between 32mm and 155mm. With extremely rapid wingbeats, these moths resemble tiny hummingbirds as they hover in front of a flower while feeding on nectar. Most are nocturnal or are active at dusk, but some very common species are diurnal. Some have wings with areas devoid of scales and resemble bumblebees in flight.

Giant Silkworm Moths – Family Saturniidae

Much sought-after by collectors, these brightly colored insects are the largest and showiest of moths. Most have wingspans of from 30mm to 150mm, but some tropical species may reach 250mm. Some display very long tail-like extensions on their hind wings. Their antennae are large and feathery, particularly among the males. The adults have only vestigial mouthparts and do not eat. They are short-lived and focus all efforts during their fleeting adulthood on mating and egg-laying. Most are nocturnal and will usually be seen fluttering around windows or outdoor lights. The name of this family comes from their copious use of silk in fashioning cocoons.

Below The elephant hawk-moth (*Deilephila elphenor*) shows off its vermilion colors, enhancing its defensive stance. One of its very large tibial spurs can be seen on its leg, and the fairly uniform antennal segments are clear. Its compound eyes look formidable, but each eye is made up of the usual ommatidia.

Tiger Moths – Family Arctiidae

The bright colors and contrasting black markings on the wings of many species in this family are reminiscent of the large cats after which they are named. Those so colored are usually day-fliers and are advertising the toxic nature of their tissues to prospective predators. At rest, the wings are held in roof-like fashion over their hairy, robust bodies. Fully extended, the wings will span between 12mm and 80mm. Like members of the Family Noctuidae, tiger moths also have a large tympanum on each side of the thorax.

Owlet Moths & Underwings – Family Noctuidae

Noctuidae, the largest family in the order, is composed mostly of drab, medium-sized moths with wingspans usually ranging between 30mm and 50mm. They resemble arrowheads when at rest, with wings folded roof-like over the body. An interesting adaptation of these moths is the presence of prominent paired tympana located on each side of the thorax. The moths themselves make no sound, but these auditory organs are tuned to the high-pitched sounds emitted by foraging bats, forewarning the moth and enabling it to execute evasive maneuvers as a bat closes in.

Above A master of camouflage and disguise, the fully-grown caterpillar of the great peacock moth (*Saturnia pyri*) of southern Europe blends with the leaves that it is eating. It is also armed with blue-topped tubercles which carry defense hairs and spines. The eight pairs of spiracles can be seen along the body, as well as the three pairs of legs.

What you can do

Trolling for moths

Females of many insect species release a type of pheromone, or chemical message, that attracts males, often over great distances. So sensitive are the finely-branched antennae of the males of certain moth species that they can detect pheromone particles in the air at a concentration of only a few parts per million and follow the increasing concentration upwind to its source.

You can lure male moths with a live female as bait in a cage fashioned from fine wire mesh. Use a species in which the males and females exhibit sexual dimorphism – different colors and/or patterns – so that they can be easily distinguished from one another. Leave the caged female outside, preferably in a light breeze, during the normal activity period of that species, and check for males at the end of it.

Females stop secreting their sex attractant after they have mated, so you will need a virgin female for this experiment.

Unfortunately, unless you rear the caterpillars yourself and separate the sexes as soon as the adults emerge, there is no way to tell if your moth is a virgin or not, so you may have to repeat the experiment several times before it succeeds.

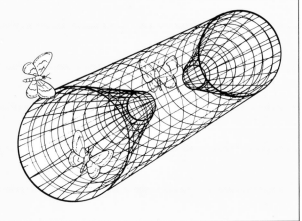

What you can do

Pinning moths & butterflies

Materials
spreading board
insect pins
broad paper strips

Mounting butterflies and moths involves a slightly more complicated procedure than that required for most insects. In order to make a good mount, the wings, critical to the identification of most species, must be properly positioned as they dry. Fully spread, the wings of butterfly and moth specimens will droop over time. To ensure that they end up on a horizontal plane, they are first positioned on a spreading board to dry at a slight upward angle.

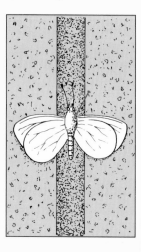

Pin the insect, freshly killed or relaxed, through the right, dorsal side of the thorax and adjust it to the correct height on the pin. Insert the pin in the groove of the board until the base of the wings is level with the top of the board at the groove. Secure the abdomen between two vertical pins to keep it immobile.

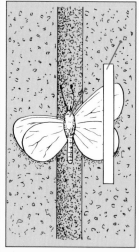

With a pin, gently push against the base of the large vein (costa), moving the front wing forward until its rear edge is perpendicular to the body. Hold it in place by applying gentle pressure to a paper strip placed over the wings, and insert a pin firmly through the paper just ahead of the wing.

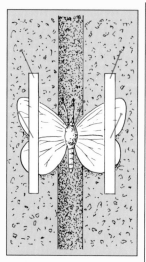

Secure the front wing temporarily with a pin through the paper immediately behind its rear edge. Repeat this procedure for the other front wing.

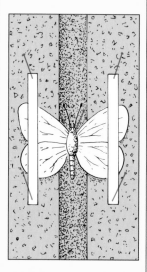

Remove the pin behind the front wing and move the hind wing forward in similar fashion until its front edge just moves under the posterior edge of the front wing. Insert a pin firmly through the paper just behind the hind wing. Repeat this procedure for the other hind wing.

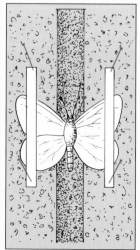

Move the antennae into their natural position and use pins to hold them. If the abdomen sags, cross two pins underneath it to form a supporting cradle.

Drying time will vary with the size of the specimen, temperature and humidity. Larger specimens may take a week or more. Keep specimens in a dry location and out of direct sunlight.

Ctenuchas & Wasp Moths – Family Ctenuchidae

Because of their diurnal habit of visiting flowers, many ctenuchid moths have adopted mimicry of dangerous species as their defense. Some are wasp-like in appearance, with slender bodies, narrow wings spanning some 28–50mm that are at least partly clear (transparent), and rapid wing-beats. Others mimic toxic insects such as tiger moths, or brilliant metallic beetles.

Clearwing Moths – Family Sesiidae

Like the Family Ctenuchidae, the day-flying clear-wing moths also mimic wasps for protection. Most have dark, slender bodies conspicuously marked with red or yellow bands. Their wings, measuring from 13mm to 60mm from tip to tip, are mostly transparent, and fringes of long hair adorn their legs. The sexes of individual species are usually colored differently.

Left The caterpillar of the bagworm moth (Psychidae family), from Penang, Malayasia, lives inside this extraordinary long bag constructed from debris spun in silk. Note the anchoring strand of silk at the side.

Top An insect with a pair of antennae three times the length of its wings, the male long-horned moth (*Adela reaumurella*) has a finely-tuned sense of smell, which it uses for detecting females and aggregating in mixed company.

Above Just a few millimeters long, this is the webworm moth (*Agonopterix arenella*, belonging to the Oecophoridae family), whose wings form a shield shape, slightly overlapping each other.

Gelechiid Moths – Family Gelechiidae

Though small, with wingspans of less than 16mm, gelechiid moths are common, and are fairly easily identified by the often brilliant metallic colors of their wings. The hind wings, which are wider than the front wings, are pointed, and have a broad hairy fringe along the outer edges.

Geometer Moths – Family Geometridae

The larval forms of these moths are the famous "inchworms" or "measuringworms" that travel with a looping motion of their bodies, pulling the rear ends up to the thoracic legs, grasping their support with posterior prolegs, and extending their bodies forward. Adults, though common, are rather nondescript with slender bodies and broad, delicate wings spanning about 8–65mm. They are most easily distinguished from similar families by their habit of resting with their wings fully spread, rather than folded.

What you can do

Insect condos

Galls are swollen masses of plant tissue caused by another organism. Examine stands of goldenrod in weedy areas during late summer. Very common on these plants are galls on the stems, where insects have laid eggs and the chemical secretions of the developing larvae have stimulated a swelling of the plant tissues around the site. If there is no hole to show where the adult has emerged, then the larvae or pupae are still inside. You can cut away galls and keep them in separate jars at home to see what emerges. Elongated, spindle-shaped galls are likely to yield gelechiid moths, while round galls will probably produce a fly from the Family Tephritidae.

Above The wavy marks on the wings of the codling moth (*Cydia pomonella*) are interrupted at the tips by a semi-metallic dusting in the shape of a thumb mark. The larvae are major pests of apples.

Above The brown house moth (*Hoffmannophila pseudospretella*, of the Oecophoridae family) is a common insect, about 10mm long, which lays its eggs in the dust and debris found in houses. Its wings are covered with silvery-gray scales.

FLIES
ORDER DIPTERA

SELECTED FAMILIES

Nobody likes them. Some are parasitic; others bite, contribute to the spread of diseases, or are simply downright annoying. So what can we say about flies that could possibly be interesting? For one thing, they are important prey for a great many animals, and for this reason alone we must tolerate them. They are also invaluable as pollinators, second only to bees and wasps, and are effective scavengers and nutrient recyclers. Since we have to live with flies, we might as well learn more about them.

Unlike the vast majority of other species, flies, of the Order Diptera, which means "two wings," have only one pair of wings. The single pair of membranous wings are attached to the mesothorax, which, enlarged and packed with flight muscles, is responsible for the high speed and wingbeat frequency of many fly species. In contrast, the prothorax and metathorax are much smaller. A fly's hind pair of wings, while not completely absent, are reduced to tiny vestigial appendages called *halteres*, which extend from the metathorax. Many families of flies have a membranous lobe at the base of each wing that covers the halteres. The presence or absence of this lobe, called a *calypter*, is often a clue to the family.

Halteres are thought to function as stabilizers in flight. Sensory organs at their bases read air currents, and the information they relay to the brain gives the fly amazing control and maneuverability, far greater than that of any other family of insects. These, combined with the claws and pads on their feet, enable them to land on virtually any surface, even ceilings!

Most adult flies have sucking mouthparts that are adapted for either sucking, sponging, or lapping liquids, although in a few species they are reduced or absent. The majority of species feed on nectar, but a significant number consume blood, plant sap, fruit juice, other insects, or decaying organic matter. Nectar feeding flies generally visit many different species of flower, but the nectar in flowers with deep corollas is inaccessible to them.

Flies engage in complete metamorphosis in the course of their life cycle. Eggs are laid individually on or near the larval food source. This is often soft, moist, decaying material, although the larvae of certain flies consume significant numbers of agricultural pests. In most species, the larvae are soft-bodied, with no legs and an indistinct head, and are commonly referred to as *maggots*. Those that are aquatic – mosquitoes and midges, for example – are much more mobile and have well-defined heads. Among the higher flies, the pupae are encased in their last larval skin, which has become a tough, impermeable capsule called a *puparium*, resistant to environmental extremes. Pupae often require precise conditions before they can develop into adults.

What you can do	
Capture techniques aerial net sweep net aspirator bait trap malaise trap (p.131) **Killing method** ethyl acetate killing jar **Preservation techniques** pin through right side of notum point (mount insect on its side)	**Note** When sweeping vegetation, slow-flying specimens can be caught with short, low strokes of the net, passing just ahead of your feet; wide strokes, passing farther ahead and just over the tops of weeds and grasses, will net some of the faster species.

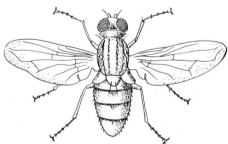

Above Midges belonging to the Chironomidae family swarm around leaves at dusk, aggregating by virtue of their pheromones, and cavorting in various prolonged dance displays. Courtship and mating eventually ensue. Their fast and furious wing-beat produces a high-pitched whine which is also probably used for aggregation purposes. Midge larvae breed mostly in freshwater.

Above Unlike most other insects flies only have one pair of wings. The reduced hind pair give the fly more maneuverability and control than any other family of insect; by acting as stabilizers that accurately read air currents.

Crane Flies – Family Tipulidae

Looking for all the world like huge mutant mosquitoes, crane flies cannot deliver the formidable bite of which they appear capable. Aside from their extremely long, stilt-like legs, they are easily distinguished from most other flies by their large size (8–65mm in length), slender body, and long and narrow wings, which protrude widely from the body when at rest. Adult crane flies are especially common in moist areas of dense vegetation, although they frequently enter buildings.

Mosquitoes – Family Culicidae

That high-pitched hum in a darkened bedroom is irritating in itself, and the anticipation of a bite dashes any hope of sleep! Most mosquitoes are less than 6mm long, but few insects are involved more directly in an adversarial relationship with humans. True, they are the vectors of some of our more serious diseases, but in temperate regions they are, for the most part, merely bothersome. Much has been made of the fact that only female mosquitoes bite, while males feed on plant juices and nectar and are important pollinators of many wildflowers. Females also consume nectar, but must have blood in order to obtain the protein necessary to nourish their developing eggs. Males are most easily recognized by their relatively feathery antennae.

Mosquitoes lay eggs in water, and many species can reproduce in almost anything that will hold water for the week or so it takes them to develop. Some actually show a preference for breeding in man-made containers. The larvae, called wrigglers, are quite active, and have a distinct head but no legs. They hang upside down at a 45-degree angle or float horizontally just under the surface and breathe through an abdominal tube that pierces the surface film. Pupae, called tumblers, are less mobile but still active swimmers, rising to the surface to breathe through two small tubes on the notum.

Horse & Deer Flies – Family Tabanidae

At least mosquitoes, with their high-pitched hum, give warning as they approach. Such is not the case with horse flies and deer flies, which fly swiftly but quietly, landing softly before inflicting a

Left The horsefly's amazing compound eyes, here seen in close-up, are remarkable for their iridescent combination of green and purple. The rows of minute ommaditia are clearly visible. The females feed on blood and then only to supply their developing eggs with protein.

Above The aquatic larva of the mosquito *Anopheles atroparvus* has several pairs of gills, through which it extracts oxygen from the water. Its body is divided clearly into head, thorax and abdomen, demonstrating these segments well. The larva defends itself by contracting and relaxing its muscles and thus wiggling away from danger.

Left Mating craneflies (*Tipula varipennis*) are a common sight in the autumn, and this pair display the main feature that makes them true flies – the halteres, or modified hind wings, read wind currents and function as stabilizers.

painful bite. Here again, it is only the females who bite, and only when they need protein for their developing eggs, while males subsist solely on nectar and pollen.

These are robust, wide-headed flies, 6–28mm long, with huge eyes that are often brilliantly colored or iridescent. The eyes of males touch one another, while those of females are separate. Horse flies are the larger of the two, and have a prominent spur on their third antennal segment. Deer flies usually have darkly-banded wings and no spurs on their antennae.

Robber Flies – Family Asilidae

Robber flies are predatory and often prey upon resting insects larger than themselves, attacking from above and using short but stout mouthparts to pierce the tough exoskeleton of their victim and drain its body fluids. Some employ a wolf-in-sheep's-clothing approach by mimicking bumble-bees, which are non-predatory. Such an appearance undoubtedly protects the robber fly as well as facilitating its advance on its prey. A few others mimic damselflies for the same advantage.

Robber flies are marked by a long, spindly

Above A male midge has a pair of large brush-like antennae which make it well equipped for detecting odors.

Right Showing its piercing mouthparts this large member of the Empididae family (*Empis tessellata*) is feeding on the flowers of buckthorn. However, most of the time they catch and kill prey by sinking their mouthparts into the bodies of their victims.

What you can do

Build a better fly trap

When they encounter an obstacle in flight, many flies will change direction and fly *upward*. You can use this item of knowledge to your advantage and construct fly traps that will collect many different specimens for you.

A classic tool of entomologists is the malaise trap, a house-shaped affair, about two yards high and the same or more long, made from the same soft mesh that you would use to make an aerial net. A large square or rectangle forms the roof, and a five-sided piece forms the side on each of the "gabled" ends. Instead of side walls, this house has one sheet of netting, three sides of which are sewn to the peak of the roof with each end hanging down the middle to form a baffle that flies will encounter. Use lead fishing sinkers at the bottom of the mesh baffle and ends to keep them vertical in a breeze. Slope the top of the roof somewhat, leaving a hole at one end into which a small jar or vial can be inserted upside down and sealed with a rubber band. It is here, at the highest point, that your captives will tend to congregate. Your malaise trap can be supported by

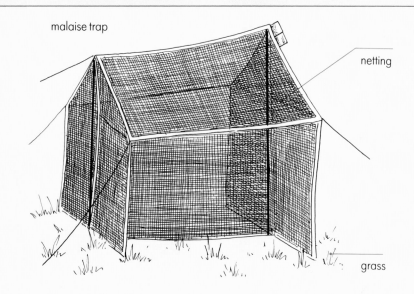

malaise trap

netting

grass

either a pole at each end or a line tied between two trees, with guy lines staked from the roof corners and peaks to hold it open.

A simpler but effective trap can be made by placing a small amount of bait in a jar lid or other such container set on the ground. Directly above this place a large white plastic funnel, upside down and supported at three points by sticks, stones, or whatever, so that it is half an inch or so above the ground. Top this with a tall, clear glass jar placed upside down over the funnel. The jar mouth must be narrower than the widest part of the funnel.

Yet another type of fly trap can be made from a half-gallon plastic milk jug. A little less than an inch above the bottom, cut several holes about half

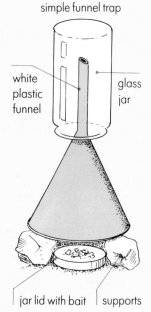

simple funnel trap

white plastic funnel

glass jar

jar lid with bait   supports

an inch in diameter. In the lid of a small screw-top glass jar, cut a hole to match the opening of the jug; seal this lid (top side down) on top of the jug with silicone bathtub caulk. Place a small amount of bait in the jug;

screw the jar onto the lid, and suspend the trap by the jug handle, hanging it from a tree branch or other support, so that it is at roughly a 45-degree angle, with the jar at the top.

The last two traps can be set with various baits. Carrion, dung, or overripe fruit will attract certain families. The recipe below for a fermented sugar bait will be attractive to moths and butterflies as well as flies.

Ingredients

1 cup low-alcohol beer
½ cup brown sugar
1 tablespoon rum
1 tablespoon honey
¼ apple finely diced and dried

Mix ingredients and let stand for 12–24 hours. Multiply ingredients 2–5 times for larger batches.

abdomen that tapers rearward, spiny legs, a bristly, bearded face, and a conspicuous concave depression on the top of the head. As flies go, they are rather on the large side, ranging between 5mm and 30mm in length.

Hover Flies – Family Syrphidae

These are also known as flower flies, both names being equally descriptive. Very common flies, they have a swift, darting flight, interspersed with frequent periods of hovering. They are quite often seen hovering over or resting on flowers, where they feed on nectar. Many strongly resemble bees or wasps in markings and behavior, but they have the distinctive head of a fly, with very large eyes that seem to occupy the entire head. Hover flies do not sting or bite, but run an effective bluff. At rest, there is a telltale nervous quivering of the abdomen. The larvae of many hover fly species are voracious predators of aphids.

Vinegar Flies – Family Drosophilidae

These are the infamous "fruit flies" that are employed in classic genetics experiments in heredity. In actuality, true fruit flies are members of another family, Tephritidae, but the name is equally appropriate to members of the Drosophilidae, because they frequent the vicinity of ripe or fermenting fruit and decaying vegetation. The larvae live in this type of medium, feeding on the yeasts found there. Adults are stocky but tiny – from 2mm to 4mm long – and a yellowish-brown in color. They, too, feed on yeasts on the surface of fruits and flowers.

Right The drone fly (*Eristalis tenax*) is a powerful fly that is known to migrate. As its name suggests, it mimics the drone honeybee in many morphological ways, including having a fairly large and shiny thorax. It belongs to the Syrphidae family, which is the largest of all fly families, and spends much of its time basking on flowers and leaves. Its larvae are carnivorous, eating aphids and other insects. Syrphids are also called hover-flies since they often hover over flowers before landing.

What you can do

See them smell

Antennae are but one of the means by which insects smell. Many have chemically sensitive *setae*, or hairs, which serve the same function. On many flies and other insects, such setae are concentrated on the tarsi, or feet.

To observe the sensitivity of these chemical sensors, mix solutions of sugar-water in concentrations of 0.5 to 10 per cent in 0.5 per cent increments (10g of sugar in 90ml of water equals 10 per cent, and so on). Capture a housefly, blow fly, flesh fly, or tachinid fly,

then keep it captive for a day or two to ensure that it builds an appetite. Stick a 1cm square of double-backed adhesive tape on the end of a small wooden dowel rod or a new, unsharpened pencil, and gently touch this to the back of the fly. Using the dowel as a handle, hold the fly about 5–6mm from the sugar solutions, starting with the weakest concentration and progressing to stronger solutions. When you reach the first solution detectable to the fly's chemical receptors, its mouthparts will dart downward in an attempt to reach the food.

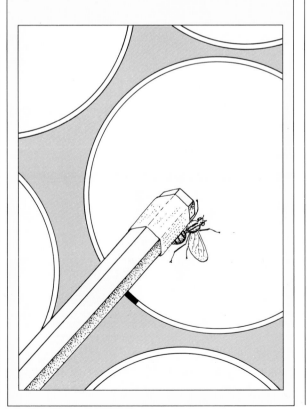

Top The March fly or St. Mark's fly (*Bibio marci*) can be very numerous in meadows and wet fields in the spring. Their larvae have primitive insect characteristics, and they feed on plant roots and decaying vegetable matter. The adults are recognizable because of their black color and the casual way in which they fly or get blown about the fields.

Above Stratiomyid flies (those belonging to the Stratiomyidae family) often have white, yellow or green markings. In this species of soldier fly (*Stratiomys potamida*) the markings are a delicate lemon yellow. Eggs are laid on vegetation close to water, and the developing larvae eat decaying vegetable matter as well as other small insects. The adults are poor fliers and frequently congregate on the umbels of umbelliferous plants.

House Flies & Kin – Family Muscidae

Possibly the best-known insect species in the world is the common housefly, *Musca domestica*, a member of the Family Muscidae. Other members of this family are much like the housefly in appearance and size (3–12mm long), and they differ from similar families, such as tachinid flies, flesh flies, and blow flies, which have stout bristles on the sides of the thorax above and below the wing bases, a feature lacking in muscid flies.

Many in the Family Muscidae feed with sponging mouthparts. A large, fleshy mass at the tip of the proboscis absorbs liquids and food particles, which are then siphoned through the proboscis. They can spread disease-causing bacteria by walking on human food, utensils, or other surfaces after having visited unsavory materials, such as manure. A few muscids have piercing-sucking mouthparts and can bite.

Blow Flies – Family Calliphoridae

Roughly the size of houseflies – about 5–15mm long – blow flies are very often metallic blue or green in color. Close inspection with a hand lens will reveal a feathery projection, the *arista*, on each of the short antennae. These are quite common flies, many of which both feed and lay their eggs on carrion or dung. A few species lay eggs in open sores, wounds, or even the nostrils of live animals (hence the name), and the resulting maggot infestation can cause the host severe irritation and weakness.

Flesh Flies – Family Sarcophagidae

Flesh flies are often identifiable by the combination of alternating black and gray longitudinal stripes on the thorax, which are absent among other flies, and a checkered abdomen. They are never metallic. Their body length ranges between 2mm to 14mm. Many lay their eggs in carrion, dung, or open wounds, although the larvae of some species are internal parasites of other insects, particularly grasshoppers and beetles.

Tachinid Flies – Family Tachinidae

Tachinid flies differ from other similar families in having a large postscutellum, a prominent lobe that occurs on each side of the thorax below the scutellum, which is the triangular sclerite behind the mesonotum. Their bodies are 3–14mm long, usually stocky, and often densely studded with stout bristles. The larvae of this large family are internal parasites of many other insects, and are second only to parasitic wasps as nature's insect population control. The eggs are either deposited directly on the host or on plant tissue which is then eaten by the host. Tachinid fly larvae usually kill their host just as they themselves become ready to emerge and pupate. Adults feed on nectar.

Above The bee fly (*Bomylius major*) is easy to identify since it has a long proboscis which it uses to probe flowers for nectar. This fly makes a high pitched whining sound, similar to that of a bee.

Above Dung flies (*Scopeuma stercorarium*) swarm on fresh dung and often mate while at least one is still feeding. Eggs are deposited on the dung, and the larvae develop rapidly.

Above Flesh flies (Sarcophagidae family) can be recognized by the strong black or gray longitudinal markings that pass down the thorax. Females incubate the eggs in their abdomens and lay live larvae. The larvae live on decaying animal or plant matter.

ANTS, BEES, WASPS, & KIN
ORDER HYMENOPTERA

SELECTED FAMILIES

The ecological significance of hymenopterans cannot be overstated. Parasitic wasps are the primary agents of insect population control. Ants aerate and fertilize the soil, recycling as many soil nutrients as do earthworms. Bees are the major pollinators of flowering plants, which in turn feed the world's terrestrial animals and provide the oxygen we breathe. In fact, about 150 crops throughout the world are pollinated mostly or entirely by bees and, as a bonus, they provide us with honey, a major crop in itself.

Entomologists consider this order to be the most advanced among insects, due to their highly specialized nature. Ants and some families of wasps and bees have evolved cooperative social hierarchies in which there is a division of labor among different castes. In such situations, workers are sterile females who care for the eggs, larvae, and pupae, maintain the nest, and repel invaders. The few males of the colony exist only to mate with the queen, the only member of the colony to lay eggs, which is her sole function.

Even though the best known hymenopterans are colonial, most species are solitary. Many of these build nests from mud or excavate nests underground or in wood or plant stems. Solitary wasps usually provision their nests with freshly killed or paralyzed insects or spiders, while non-colonial bees stock theirs with pollen and nectar. It is thought that, even though the parents and offspring never see one another, provisioning by the parents is the precursor to true social behavior, in which the adults of one generation cooperatively feed and care for immature insects of the next.

Some other solitary species do not build nests, but lay their eggs in or on the eggs, larvae, or pupae of other insects. These are generally termed parasitic, although they actually kill their only host. They can more correctly be called *parasitoids*, for they fall somewhere between true parasites, which do not kill their host, and predators, which kill more than one victim. If the immature wasp feeds from the outside of its host, the adult female will first sting and paralyze the victim before depositing her egg.

bee

ant

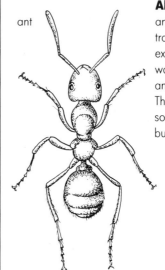

Above and **left** Both bees and ants have two pairs of transparent wings (with the exception of the wingless worker ants), "wasp waists" and chewing mouthparts. The best known species are social and live in colonies but the majority are solitary.

Probably the main factor in the evolution of most hymenopteran lifestyles was the development of the "wasp waist." In all but the more primitive families, namely sawflies and horntails, the first abdominal segment, or *propodeum*, is separated from the rest of the abdomen by a highly flexible and often constricted hinged joint, the *pedicel*, which allows precise movement of the abdomen in egg-laying and defense, and also permits the insects to turn around in the tight confines of a burrow or nest.

Other features that distinguish members of the Order Hymenoptera are their chewing mouth-parts, 5-segmented tarsi, and four membranous wings (when present). The forewings are somewhat larger than the hind wings, and both pairs generally have few veins and large cells, the spaces bordered by veins. Many have chewing-lapping mouthparts, in which the labium and maxillae are modified into a tongue-like apparatus for drinking liquids. All species undergo complete metamorphosis. Among those bees, wasps, and ants that sting, the stinger is actually a modified ovipositor, or egg-laying structure. Sawflies and horntails, despite their ominous appearance, do not sting.

What you can do

Capture techniques	Preservation techniques
aerial net	
sweep net	pin through right side of
aspirator	notum point
bait	70 per cent alcohol vial

Killing method
ethyl acetate killing jar
immersion in 70 per cent
alcohol

Note Many hymenopterans can sting, and severe allergic reactions may occur in sensitized individuals. Handle with care!

Left It is comparatively easy to determine whether ants, bees and wasps belong to the three basic groups of hymenopterans. However, it is much more difficult to place the solitary bees and solitary wasps within their individual groups. In addition, there are true flies that mimic bees and wasps, further confusing the matter. This is the wall mason wasp (*Ancistrocerus parietinus*), which is a British solitary wasp.

Sawflies – Family Tenthredinidae

Tenthredinidae, the largest family of sawflies, includes most of the species commonly encountered. Black or brown, and between 3mm and 20mm long, they are most easily distinguished from other sawflies by their thread-like antennae, which have from 7 to 10, but usually 9, segments. As with all other sawflies and horntails, the junction of abdomen and thorax is broad, unlike the more familiar families of hymenopterans. Females have a saw-like ovipositor with which they insert eggs into plant tissues, and they are often colored differently from males of the same species.

Females of many sawfly species, as they lay an egg, inject a substance into the plant that stimulates gall formation, a proliferation of plant cells in the tissues around the egg. This serves as both shelter and as a protein-rich food source for the larva.

Horntails – Family Siricidae

The absence of a "waist" gives horntails a cylindrical appearance, but otherwise they are rather wasp-like. Their bodies are usually from 20mm to 40mm long, brown or black, and often marked with yellow. The name stems from a long, spiny projection on the last abdominal segment of both males and females. Beneath this, females are equipped with a long, stout ovipositor.

Ichneumon Wasps – Family Ichneumonidae

Ichneumons constitute a large family of parasitic wasps embracing a great variety of sizes and colors. Most are uniformly colored, from black to yellow, but some are boldly patterned or have conspicuous white or yellow segments in the middle of their long antennae. The antennae have more than 15 segments and are at least half as long as the body. These are slender insects, often with abdomens that are quite narrow at the base and expand steadily rearward. Many females trail long, thread-like ovipositors behind them, some of which have taste receptors to help them identify suitable hosts. Ichneumons resemble the smaller, stockier parasitic wasps of the Family Braconidae. The larvae of both families are major parasites of other juvenile insects, especially those in the Order Lepidoptera, although they also attack larvae of many beetles and of their own order.

Above Sawfly larvae are gregarious, feed voraciously and, in the case of this gooseberry sawfly (*Pteronidea ribesii*) are commercial pests.

Above With her long ovipositor, (seen here between her second and third pair of legs) buried deep in the wood, a female of the greater horntail (*Urocerus gigas*) furnishes each hole with an egg.

Above Using a combination of vibrations in the wood and her antennae to sense them, the female of the parasitic wasp, *Rhysella approximator*, positions her ovipositor in the correct place to pierce the larvae of the woodwasp (*Xiphydria camelus*) and lay her eggs in them.

Left The ichneumon wasp (*Pimpla instigator*) is principally a parasite of butterflies. The wasp lays its eggs through the cuticle of the caterpillar, and the wasp's larvae feed on the non-vital parts of the body without killing their host. When fully grown, they simply eat their way out of their host and pupate on the writhing remains.

139

Velvet Ants – Family Mutillidae

Despite their name, these are not ants at all, but very hairy parasitic wasps. They differ from true ants in that they are covered with dense, brightly colored hair, their antennae are not elbowed, and the pedicel at the base of the abdomen is broader than that of an ant and lacks a dorsal hump. Most have blazing red, yellow, or orange patterns and are from 6mm to 25mm long. Females are wingless, smaller than males, and can deliver a very painful sting. Their larvae are external parasites, mostly upon the larvae and pupae of bees and other wasps.

Ants – Family Formicidae

"The ants go marching one by one. Hurrah! Hurrah!..." begins the children's song, reflecting the air of order and cooperation that is synonymous with ants. All ants are social insects, living in colonies with well-defined castes and divisions of labor. Most colonies are composed of a large worker caste of wingless, sterile females and a reproductive caste of winged, fertile males and females, the latter also known as queens.

Workers are aptly-named, for they perform every function critical to the survival of the colony except for reproduction. Building and maintaining the nest, repelling intruders, gathering food, and tending eggs, larvae, pupae, and the queen are all tasks which fall to the workers. Most ant species are scavengers, but many will harvest leaves and seeds, raise fungi in subterranean chambers, prey upon other invertebrates, or herd and protect aphids while harvesting their honeydew secretions.

An ant nest is a maze of tunnels underground or in decaying wood. Workers make regular forays in search of food, and in order to find their way to a recently discovered food source or back to the colony, they lay down a scent trail as they go. In addition, each colony has its own distinct odor, which largely determines recognition among members and rejection of outsiders. In spring and fall, some members of the reproductive caste leave the colony in a short mating flight. After mating, males die and females lose their wings and usually begin a new colony, each caring for the first brood of workers by herself.

Ants are usually black, brown, or reddish, and they range between 1mm and 25mm in length. Their distinguishing features are unmistakably elbowed antennae and a slender pedicel between the thorax and abdomen that, when viewed from the side, reveals one or two distinct humps. Many species are able to spray formic acid or other noxious chemicals as a means of defense.

Vespid Wasps – Family Vespidae

Vespid wasps include the familiar yellow jackets, hornets, and paper wasps, among others. They are from 10mm to 30mm long and have conspicuously notched eyes. Most are brown or black, and some species have distinct white or yellow markings. Their wings have a pleated appearance when folded over the back at rest, a trait that distinguishes them from the similar sphecid and spider wasps. Many vespids are colonial and lay eggs in combs of hexagonal cells made of a papery substance. Such nests may be open or enclosed

Top Nests of the *Polistes* wasp, from southern Europe, are never very large, are constructed out of wood fiber and are found on buildings, in stone walls or, rarely, in vegetation.

Above A freshly-emerged specimen of the common wasp (*Vespula vulgaris*), complete with all its body hair, looks out of the hive and defies any robbers which might fancy mounting an invasion.

with only one entrance near the bottom. Other vespids are solitary, and lay eggs in underground cells or cells constructed of mud. The cells are provisioned with chewed insect prey rather than whole insects. Yellow jackets and hornets employ the same sort of caste system as do ants and bees, each colony having only one fertile female and a large caste of infertile female workers, whereas colonies of paper wasps consist of males and a group of fertile females living cooperatively in the same nest. All species in this family can inflict painful stings, and alarm pheromones released by disturbed individuals can incite gang attacks.

Spider Wasps – Family Pompilidae

These solitary wasps are typically seen running across the ground while nervously flicking their dark or amber-colored wings. Many are a shiny bluish-black color or have red or yellow markings. They measure from 10mm to 50mm in length and have relatively long legs, with the hind femurs extending to the tip of the abdomen. Spider wasps take their name from the females' habit of provisioning their underground nests with paralyzed spiders, on which the eggs are laid. Despite this habit, the adults usually feed on nectar.

Sphecid Wasps – Family Sphecidae

Sphecid wasps are solitary, and are characterized by a short, collar-like pronotum, with a lobe projecting backward on each side. They are from 10mm to 55mm in length and, unlike those of the similar spider and vespid wasps, their wings do not fold over the abdomen when at rest. The nests, provisioned by females with insects or spiders, are usually either mud cells attached to buildings or

Left A mud-daubing wasp (*Sceliphron formosum*) collects mud from a tiny puddle ready to make her nest of mud on a ceiling or overhang. Small rows of mud nests are made together, and each is furnished with pollen and an egg.

Above The ingenious spider hunting wasp (*Anoplius infuscatus*) outwits the spider and paralyzes it, prior to taking it away to furnish its nest, where it will place an egg on the living food supply. The wasp is a member of the Pompilidae family.

What you can do

Ant farming

Materials for alternative design

2 pieces glass or plexiglass, 12in × 13in (30.5 × 33cm)
4 pieces wood, 12in × ½in × ½in (30.5 × 1.5 × 1.5cm)
1 piece wood, 4in × 14in × ½in (10 × 35.5 × 1.5cm)
small wood screws
2 large rubber bands
funnel
camelhair artist's brush
cotton
fine sandpaper
several vials
garden trowel
large brown paper envelope

Granted, keeping an ant farm does have a certain childish connotation, but it's really not just an occupation for children. You can maintain a thriving ant colony inside your own home and observe them as they go about their daily lives in the confines of a self-contained system.

One of the simplest ant farms to build requires two straight-walled glass containers, one slightly smaller than the other so that it nests inside the larger vessel with a gap of about ½in between the two around the entire circumference. The inner jar should be slightly shorter, so that the outer one can be capped.

1 Use a bit of modeling clay or similar adhesive substance to hold the inner jar precisely in place while you carefully fill the gap between the two with loose, very sandy soil that has been sifted to remove stones and other debris.

2 Bring the soil level to just below the top of the inner jar and tamp it down gently with the eraser end of a pencil.

3 Food and a piece of damp sponge for moisture can be added to the inner jar. After introducing ants to the formicarium, cover the mouth of the outer jar with fine nylon mesh held in place by a rubber band.

Alternative design

1 Another successful design requires that you first construct a U-shaped frame from three 12in lengths of wood that are each ½in square in cross section.

2 Place the bottom ends of the two pieces forming the sides of the U on *top* of the bottom piece, *not* aside of it.

3 Sandwich this between the two sheets of glass or clear plexiglass with the excess extending past the open end of the U. Any sharp edges should be sanded first. Plexiglass can be drilled and attached with wood screws, but glass is best secured with two heavy rubber bands.

4 Drill a ½in hole in each side of the U, centered about 1½in below the top of the wood, and plug these holes with wads of cotton.

5 Mount the frame, open end up, in the center of a 4 × 14in wooden base, for stability.

6 Fill with soil as described above, tamping it down periodically, until it is about level with the bottom of the holes.

7 Finally, attach a fourth length of wood, with the same dimensions as the other three sides, between the glass or plexiglass on top, and drill a ½in hole in the middle. This design lends itself well to expansion, as you can link two or more ant farms by running clear plastic tubing, available from aquarium supply stores, from one side hole to another.

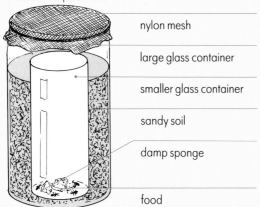

nylon mesh
large glass container
smaller glass container
sandy soil
damp sponge
food

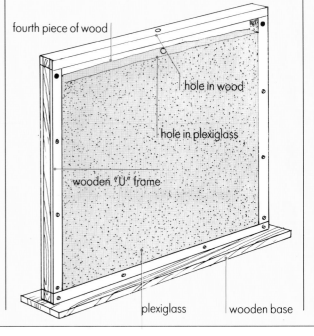

fourth piece of wood
hole in wood
hole in plexiglass
wooden "U" frame
plexiglass
wooden base

Finding the ants

Try to get your ants and soil from the same vicinity. Look for ant colonies under large, flat rocks in vacant lots, meadows, or other areas not recently disturbed. Bait several vials with honey and lay them open, near to the colony. When you have about a hundred ants in the vials, plug the openings with wads of cotton. Next, unearth the colony with a garden trowel, placing each scoop of soil on a large sheet of white paper and sifting through it until you find the queen, who will be much larger than the rest. Put her in a separate vial.

Putting ants into the farm

First place a dinner plate upside down on a large tray. Pour enough water in the tray to form an island, from which the ants cannot escape. Place the ant farm on the island and put a small funnel in the top hole. Coax the queen from her vial and down the funnel with a camel hair artist's brush. Introduce the rest of the ants in similar fashion, and plug the hole with a wad of cotton. Those that climb out of the funnel can be trapped again by the same method as before.

The cotton stoppers will allow a free flow of air into the ant farm. Your ants will also need water, which can be supplied with a bit of damp sponge every few days. Since most ants are scavengers, you should have no trouble finding foods they like. Bread and cracker crumbs, bits of fruit or ground meat, and honey will usually be taken readily, but be careful not to overfeed. Also, since ants prefer darkness in the nest, cover the ant farm with a large, dark envelope or other material to keep them content, and try to use artificial light while observing them.

To retrieve pieces of sponge and unwanted food with greater ease, you may wish to establish the feeding area in a screw-top glass jar, connected to the farm with a piece of clear plastic tubing. Once the nest is established and the queen has begun laying eggs, the ants will consider it home and return to it freely. You can run a plastic tube from the farm and out of a window to the ground, allowing them to come and go as they wish, gathering their own food outdoors.

Your ant farm will lend itself to many interesting experiments. Observe the different reactions when you remove a few ants from the colony and later return them, compared to the colony's response when

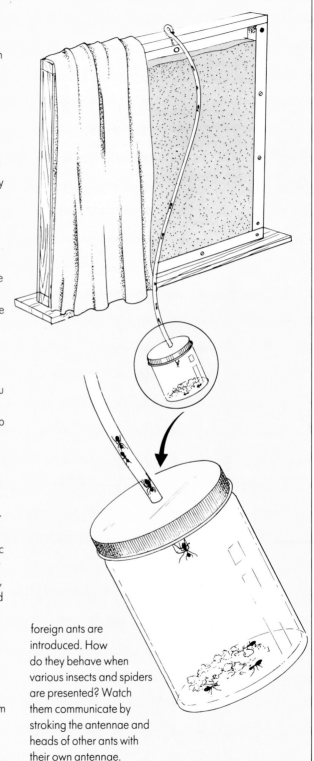

foreign ants are introduced. How do they behave when various insects and spiders are presented? Watch them communicate by stroking the antennae and heads of other ants with their own antennae.

rocks, or are found underground or in natural cavities. Some sphecid wasps exhibit an elementary form of social behavior, in which small groups of females cooperate in constructing a mud nest, although each provisions her own cells. This behavior is thought to be the immediate precursor to true insect social organizations.

Bees – Superfamily Apoidea

Because of their many similarities, all families of bees may be conveniently considered in one superfamily, which consists of the Families Colletidae, Halictidae, Andrenidae, Melittidae, Megachilidae, Anthophoridae, and Apidae. These are separated primarily on the basis of differences in mouthparts and wing vein patterns. By far the most significant is Apidae, which includes bumblebees, carpenter bees, and honeybees. Honeybees and bumblebees are the only social bees. Others are solitary, and most build nesting tunnels either underground or in wood or plant stems, provisioning them with nectar and pollen balls, on which the eggs are laid. Bees range from 4mm to 25mm in length and are often black or brown, with white, yellow, gold, or orange bands.

Bees are highly specialized for gathering nectar and pollen. The long proboscis is ideally adapted for probing deep into flowers for nectar, and pollen sticks to their hairy bodies as they brush against the anthers of flowers. Pollen brushes and combs – stiff hairs located on the hind legs of females – are used to comb pollen from the body hair and transfer it to pollen baskets, located on the hind tibiae. Pollen baskets sometimes become so overloaded that the bees have difficulty flying. Coincidentally, bees are the primary pollinators in the world, and many plants are completely dependent upon them for this service.

Left The leaf-cutter bee (*Megachile willoughbiella*) methodically cuts out a precise section of leaf, flying off with this held between the legs. The bee makes a nest-cell from the curled piece of cut leaf, which is then furnished with a ball of pollen on which the eggs are laid. In structure and color the leaf-cutter bee is very similar to a honeybee, and its close relation is the mason bee.

Above The tawny-mining bee (*Andrena armata*) is named after its abdomen and its mining activities in the ground. Hatching in its subterranean cell during the winter, the bee comes to the surface and flies off in the spring. Some members of the Andrenidae family live in colonies or "villages" which may number a thousand or more nests.

Left The carpenter bee (*Xylocopa violacea*) is very common in southern Europe and visits many kinds of flowers, especially those of wisteria. They provision holes in buildings and walls with pollen and lay eggs on top of it. Carpenter bees will sometimes become drunk on fermented nectar and fall to the ground completely exhausted.

What you can do

Training bees and wasps

Learning by association

1 On an outdoor table in summer, lay out a large sheet of white paper and, using a black marker, draw a number of bold shapes or patterns on them; a star, square, triangle, circle, X, radiating lines, and so on, but only one of each. Cover this with a sheet of clear plastic to protect it against rain or dew.

2 Place a small cup or bowl by each shape, and fill all but one with water. Fill the last one with a 25 per cent sugar solution, made by dissolving one part sugar in three parts boiling water and allowing it to cool. Refill the bowls as

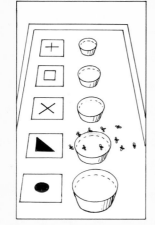

needed for about two weeks, but always keep the sugar solution by the same shape. If the sugar water is by, say, the triangle, and at the end of two weeks you move it to another location, you will find that most of the visiting bees and wasps will still tend to congregate at the triangle.

A sense of time

1 Place the sugar solution in the same location at exactly the same time every day for a one hour interval. Do this for two or three weeks, and you will find that for some time afterwards the bees will still show up at the appointed time, even if you do not bring the food.

2 You can even verify that they are the same individuals by placing dabs of bright paint or nail polish on the back of each bee's thorax as she drinks, taking care not to get any on her wings.

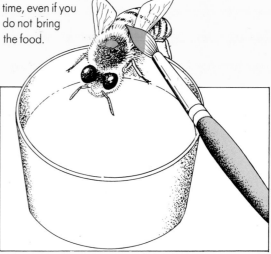

MINOR INSECT ORDERS

SELECTED ORDERS

The orders of Class Insecta covered have been the "Big Eight," which, combined, include about seven of every eight insect species. The remaining 16 or so orders in this class, though they include some rather significant and familiar families, are unceremoniously known as "minor" insect orders. In keeping with the practical nature of this book, most of it has been devoted to those orders you are most likely to encounter. However, there are no insignificant organisms, and though it is not intended as a field guide, *The Practical Entomologist* would be incomplete without at least mentioning a few of the smaller orders. Orders of note that are not described here include bristletails, springtails, termites, earwigs, chewing lice, sucking lice, scorpionflies, and fleas.

Top Differences in two castes of the termite (*Macrotermes gilvus*) are shown here; compare the large soldier with its powerful jaws and enlarged head, with the smaller worker.
Center A pair of common earwigs (*Forficula auricularia*) can be identified by the shape of their forceps, for those of the male are strongly curved, while those of the female are not. Wings are present, but rarely used.
Right Only a few millimeters long, the water springtail (*Podura aquatica*) relies on using surface tension for moving about.

Mayflies – Order Ephemeroptera

The order name means "to live but a day," which refers to the lifespan of its adults. Though some species may persist as adults for several days, most will die within 24 hours of molting into their winged form. These delicate insects are ill-equipped in the final stage of their life cycle to do anything but engage in frenzied mating swarms over the waters in which they lived as naiads. Large numbers of adults typically mature at once in a synchronized "hatch." Males seize any female that ventures into their swarm, copulate in flight, and die immediately afterward. Females enjoy a slightly longer life, submerging once again and laying eggs attached to underwater objects by short filaments. This happens within an hour or so of mating, and the females expire shortly afterwards.

Adult mayflies are yellowish or brownish, with an upward tilt to their long, slender abdomens. Most noticeable are the two or three very long, filamentous appendages extending from the tip of the abdomen. Also conspicuous are their many-veined, triangular forewings and roundish hind wings, which are held vertically over the back at rest. Mayflies, despite their name, are commonly seen in both spring and summer, and their presence is an indicator of a relatively unpolluted body of water. There are three families in this order: Ephemeridae, Heptageniidae, and Baetidae.

Cockroaches – Order Blattodea

The scourge of homeowners everywhere, cockroaches are not quite as bad as their reputation. While they may contaminate improperly stored food with their feces and produce disagreeable odors, they do not transmit human diseases. They forage at night and hide by day, and though winged, they usually escape by running. Intense campaigns to eradicate them have produced generations of survivors immune to most pesticides. Through such resilience, they have become one of the oldest types of winged insect, dating back more than 350 million years.

Cockroaches are brown, with bodies up to 60mm long that are flattened, helping them slip into narrow cracks to hide. Their pronotum extends far forward to produce a hooded appearance, and they have long, swept-back, thread-like antennae that help them to feel their way in the dark. Most species prefer warm, damp, subtropical

Top The naiad of the mayfly (*Ephemera danica*) crawls from the water and secures itself on a stem prior to the adult emerging. The aquatic immature stage may last three years in some species, but the ephemeral nature of the adult means that it is usually dead within a day.

Above Cockroaches are opportunists and scavengers that will eat all sorts of vegetable remains. Here Australian cockroaches (*Periplaneta australasiae*) devour an apple. Being very alert they scurry off when disturbed and have sharp spines on their legs to resist attack.

conditions, but representatives occur almost everywhere there is human habitation. Once considered part of the Order Orthoptera, they are now a separate order consisting of Families Blattidae and Blattellidae.

Mantids – Order Mantodea

Mantids are easily recognized by their long (up to 150mm), slender bodies and slow but graceful movements. They have two pairs of walking legs, and toothed, muscular forelegs that are well adapted for seizing prey with amazing swiftness. Typically held close to the body in a praying position, this front pair of legs has inspired the common name "praying mantis." Mantids employ detailed mimicry of green plant stems and leaves to stalk or ambush their victims.

The head has an amazing range of motion, so much so that the insect projects a quite human image as it looks over its shoulder or cocks its head to get a better view. Powerful mouthparts enable mantids to cut through the tough armor of most insects, their primary prey. Because of their predatory habits and voracious appetites, they are extremely beneficial, especially to gardeners. Females lay hundreds of eggs in a foamy mass that hardens into a tough, insulating case. Egg cases are usually attached to branches or man-made structures, but the mantids themselves can be found in dense vegetation. These, too, were once considered part of the Order Orthoptera.

Walkingsticks & Timemas – Order Phasmatodea

Masters of mimicry, walkingsticks and timemas are nearly invisible in their natural habitats. Walkingsticks, brown and up to 150mm long, are bizarrely modified to resemble leafless twigs. They even adopt a slow, swaying stride that copies the motion of wind-blown vegetation. Timemas resemble the leaves themselves. Unlike mantids, however, the camouflage of walkingsticks and timemas is purely defensive. Their metamorphosis is incomplete, and they can reproduce by parthenogenesis. The two families are Phasmidae and Timemidae.

Stoneflies – Order Plecoptera

The order name, which means "folded wings," refers to the broader hind pair, which is folded fan-like and covers most or all of the abdomen of individuals. The bodies, which are between 6mm and 64mm in length, are flattened, mirroring the streamlined nature of the naiads, which inhabit swiftly-flowing streams. Fairly long antennae and a pair of usually long cerci on the tip of the abdomen are also hallmarks of this order.

Another indicator of relatively pure water, stonefly-naiads are predators of well-aerated streams and rivers. Their brief adulthood is spent crawling around rocks near the water, hence the common name. Adults are capable only of a weak, fluttering flight. Families include Nemouridae, Capniidae, Taeniopterygidae, Peltoperlidae, Leuctridae, Pteronarcyidae, Perlodidae, Perlidae, and Chloroperlidae.

Dobsonflies, Alderflies, & Fishflies – Order Megaloptera

Because of their densely-veined wings, these insects were once classified in the Order Neuroptera, the net-veined insects, and even now the chief difference between the two orders is a matter of wing venation. Megaloptera, which means "ample wings," refers to the large hind wings. At rest, these are folded, roof-like, over the abdomen. Despite the large wings, adults are weak fliers, and are often seen fluttering near lights. Adults are between 9mm and 70mm in length, with long, swept-back antennae. They have chewing mouth-

Left Praying mantids are masters of disguise; often leaf-shaped, they may be colored green, brown or pink, sometimes varying as they grow older. They often lie by flowers waiting for visiting insects to arrive, whereupon they strike out with their battery of spines and impale the insect on their modified first pair of legs. They then start to eat the prey, dead or alive. This is a female praying mantis (*Galinthias amoena*), from Kenya, eating a fly.

Top The giant ant-lion (*Palpares weelei* of the Myrmeleontidae family), from the tropical dry forest of Madagascar, has large, spatula-shaped wings and relatively short antennae.

Right Ant-lion larvae live in groups in sandy soil where they each lie in wait at the base of a pit, waiting for unsuspecting prey to fall in. The sides of the sand pit are relatively steep, and any insect that attempts to cross the sand will fall directly into their waiting jaws.

Above This fabulous day-flying member of the lacewing family (Neuroptera) is found in south-western Turkey and is called *Neuroptera sinuata*. It has evolved colors and patterns similar to those of moths or butterflies.

parts, and the mandibles of male dobsonflies are massive and tusk-like, about three times longer than the head. Larvae are voracious aquatic predators with an appetite for black fly larvae. Metamorphosis is complete, with the larvae leaving their ponds or streams to pupate. Families include Corydalidae and Sialidae.

Net-veined Insects – Order Neuroptera

The order name means "nerve wings," a reference to the many-branched veins in their wings, which are all about equal in size and shape and are held roof-like over the abdomen when at rest. Like the Megaloptera, they have a weak, fluttering flight. Most adults are slender and delicate, measuring from 6mm to 45mm in length, with large compound eyes and long, slender antennae. Metamorphosis is complete, and the larvae of most species are very beneficial, preying on many

destructive insects, such as aphids. Families include Coniopterygidae, Chrysopidae, Hemerobiidae, Sisyridae, Mantispidae, Myrmeleontidae, Ascalaphidae, Polystoechotidae, Dilaridae, and Ithonidae.

Caddisflies – Order Trichoptera

Caddisflies are notable for the aquatic larvae's habit, many species possess, of building portable protective cases around themselves by cementing together bits of sand or debris. Adults are quite moth-like, but lack the wing scales and coiled proboscis of an adult lepidopteran, having instead chewing mouthparts and fine hair on their wings. At rest, the wings are held roof-like over the abdomen. Most species are nocturnal, often fluttering around lights and hiding during daylight hours. Adults are between 7mm and 25mm long and gray or brown in color.

Above While some booklice are found on books, to which they are attracted by the vegetable content of fibers and glues, other booklice live in the open among leaf litter and debris. This winged psocid (*Psococerastis gibbosa*) has a marvellous violet sheen. Many other species of psocid, particularly those found in libraries, are wingless.

Right Caddis flies have aquatic larvae which are easy to find on the undersides of rocks. The larvae spin a fine web with their own silk and use this to filter food debris from the water of the stream. They also use tiny stones in the construction of their caddis cases, and the design and shape of a case can be a help in identifying the species.

GLOSSARY

This oil beetle (*Lydus syriacus*) clearly illustrates the three body parts: **head, thorax** and **abdomen**.

This oil beetle (*Lydus syriacus*) clearly illustrates the three body parts: **head, thorax** and **abdomen**.

abdomen: the posterior one of the three major subdivisions of an insect's body.

anal: toward the posterior end or side.

antennae: paired sensory appendages, one located on each side of an insect's head.

anterior: toward the head.

asexual reproduction: the production of viable offspring without the fertilization of eggs by sperm.

calypter: a lobe on either the posterior wing margin or close to the base of the wing on the thorax of certain fly families.

carrion: decaying flesh.

caste: a specialized segment of an insect colony, with very distinct functions.

cell: an area of wing membrane partially or completely bounded by veins.

cephalothorax: the first subdivision of a spider's body, composed of the head and thorax.

cerci: paired sensory appendages located at the posterior of the abdomen on some insects.

chitin: a component of the exoskeleton that is very tough, flexible, and has some degree of resistance to most chemicals.

chrysalis: a butterfly pupa not encased in a cocoon.

clubbed antennae: those that are abruptly enlarged at the tip.

cocoon: composed of silk fibers secreted by a larva, this is the protective case in which the larva will pupate.

colony: a congregation of a single species of insects, living together cooperatively.

compound eyes: a pair of visual organs, one on each side of an insect's head, each composed of few to several thousand photoreceptive units radiating outward and terminating in a lens, all of which are joined to form facets of the eye.

coxa: the basal segment of an insect's leg.

cuticle: the non-cellular, secreted outer covering of an insect's body.

dimorphism: the occurence of two distinct forms of the same species.

dorsal: pertaining to the back of an organism, usually the upper surface.

elbowed antennae: those bent at a 90 degree angle near the middle.

elytrum(a): the hardened or leathery forewing(s) of beetles.

epidermis: the layer of living cells that underlie and secrete the insect's cuticle.

exoskeleton: an external, waterproof, protective body covering, composed of chitin, sclerotin, and waxes, which houses and supports the internal organs, muscles, and other tissues.

eyespots: protective markings on the wings of certain insects, intended to mimic large vertebrate eyes.

facet: the external surface of an individual unit of a compound eye.

femur: the third segment, counting from the body, of an insect's leg, found between the tibia and trochanter.

Cockchafer beetle (*Melolontha melolontha*) showing flabellate **antennae**.

haltere: the vestigial hind wing of a fly, it functions as a stabilizer in flight.

hemelytra: distinctly thickened or leathery fore-wings of true bugs, which have overlapping membranous tips.

hemimetabolous development: incomplete metamorphosis, in which immature nymphs resemble adults of the species as soon as they hatch from the eggs. Their size increases with successive molts, as do the development of wings and sexual organs. Aquatic hemimetabolous insects in their immature form are called naiads.

holometabolous development: complete metamorphosis, in which the insect passes through four distinct stages in its life cycle (egg, larva, pupa, and adult), and the immature stages do not resemble the adult.

honeydew: a sweet excretion of many true bugs, especially aphids.

instar: a stage of insect development between molts. The first instar is the stage between hatching from the egg and the first molt.

invertebrate: an animal without a backbone.

labium: the most posterior of an insect's mouthparts, analogous to a lower lip.

labrum: the most anterior of an insect's mouthparts, analogous to an upper lip.

larva: the second stage in a holometabolous insect's life cycle, between the egg and pupa.

mandibles: a pair of insect mouthparts, analogous to jaws and normally used for chewing.

margin: an edge of an insect's wing.

maxillae: a pair of insect mouthparts, usually posterior to the mandibles and used for food

Lycus constrictus beetle showing the hardened, leathery **elytra**.

Pupa of a large white butterfly (*Pieris brassicae*).

handling.

mesonotum: the dorsal surface of the mesothorax.

mesosternum: the ventral surface of the mesothorax.

mesothorax: the middle segment of an insect's thorax, which bears the middle pair of legs and the forewings, if present.

metamorphosis: the transformation of a larva, nymph, or naiad into an adult insect.

metanotum: the dorsal surface of the metathorax.

metasternum: the ventral surface of the metathorax.

metathorax: the third, or most posterior, segment of an insect's thorax, which bears the third pair of legs and the hind wings if present.

molt: the shedding of the rigid cuticle to allow for growth or metamorphosis.

naiad: the immature form of an aquatic hemimetabolous insect.

notum: the dorsal surface of the thorax.

nymph: the immature form of a terrestrial hemimetabolous insect.

ocelli: simple eyes with a single lens that can detect varying degrees of light intensity but most probably cannot discern images.

ovipositor: the egg-laying appendage located at the posterior end of female insects.

palpi: paired sensory appendages of mouthparts.

parthenogenesis: asexual reproduction, in which eggs develop without being fertilized by sperm.

pedicel: the slender stalk of the abdomen found

on most members of the Order Hymenoptera.

pheromone: chemical messages emitted by insects.

pleuron: the side of an insect's abdomen.

posterior: away from the head.

proboscis: elongated, tube-like mouthparts, adapted for consuming liquids.

prolegs: soft, unjointed, paired appendages used for support and typically found on caterpillars.

pronotum: the dorsal surface of the prothorax.

prosternum: the ventral surface of the prothorax.

prothorax: the first, or most anterior, segment of the thorax, to which the first, or most anterior, pair of legs are attached.

pupa: the third, inactive stage of a holometabolous insect's life cycle, between the larval and adult stages, in which the insect transforms into an adult.

pupate: the process of becoming a pupa.

raptorial legs: thick, muscular forelegs of predatory insects, used to grasp and hold their prey.

sclerite: one of the hardened plates that compose the exoskeleton.

scutellum: a triangular sclerite of the mesonotum, located between the forewings and especially prominent among beetles and true bugs.

segment: a section of an insect's body or appendage, bordered by flexible membranous regions.

seta(e): hair-like sensory structure(s) that may be adapted to detect touch, smell, taste, or sound.

social insects: those species that live cooperatively in colonies and exhibit a division of labor among distinct castes.

The **scutellum** can easily be identified on these squashbugs (*Sagotylus confluentus*).

Spotted longhorn beetle (*Strangalia maculata*) clearly showing the various **leg segments**.

solitary insects: non-social insects that do not form long-term associations with other members of their species.

spiracle: a breathing hole, usually on the abdominal surface, through which gases enter and leave an insect's respiratory system.

sternum: the ventral surface of an insect's thorax and abdomen.

tarsus: the terminal section of an insect's leg, composed of one to five segments and attached to the tibia.

tergum: the dorsal surface of an insect's abdomen.

thorax: the middle of the three major subdivisions of an insect body, between the head and abdomen, to which all legs and wings are attached.

tibia: the fourth segment of an insect's leg, between the femur and tarsus.

trachea: part of the respiratory system, a tube through which gases move to and from internal organs.

trochanter: the second segment of an insect's leg, between the coxa and femur.

tympanum: an auditory organ consisting of a vibration-sensitive membrane on the abdomen or forelegs of grasshoppers, cicadas, and some moths.

ventral: pertaining to the side of the body opposite the back, usually the lower surface.

CONTACTS

Biological Equipment and Supplies

American Biological Supply Co.
1330 Dillon Heights Ave.
Baltimore, MD 21228
(301) 747–1797

Carolina Biological Supply Co.
2700 York Road
Burlington, NC 27215
(800) 547–1733

Connecticut Valley Biological Supply Co.
82 Valley Road
South Hampton, MA 01073
(413) 527–4030

Edmund Scientific
101 East Gloucester Pike
Barrington, NJ 08007
(609) 573–6250

Fisher Scientific
52 Fadem Road
Springfield, NJ 07081
(800) 441–2374

Frey Scientific Company
905 Hickory Lane
Mansfield, OH 44905
(216) 589–1900

Nova Scientific Corporation
111 Tucker St.
P.O. Box 500
Burlington, NC 27215
(919) 584–0381

Ward's Natural Science Establishment
5100 West Henrietta Road
Box 92912
Rochester, NY 14692
(800) 962–2660

Watkins and Doncaster
Conghurst Lane
Four Throws
Hawkhurst
Kent TN18 5ED, UK
05805–3133

Worldwide Butterflies Ltd
Compton House
Nr. Sherbourne
Dorset DT4 QN, UK
0935–74608

Entomological Organizations

Butterfly World
Tradewinds Park
3600 West Sample Rd.
Coconut Creek, FL 33073
(305) 977–4400

Calloway Gardens Day Butterfly Center
Pine Mountain, GA 31822
(800) 282–8181

Hole-in-Hand Butterfly Farm
Hazleton, PA 18201

Lepidoptera Research Foundation
Santa Barbara Museum of Natural History
2559 Puesta del Sol Road
Santa Barbara, CA 93105

Lepidopterists' Society
Department of Biology
University of Louisville
Louisville, KY 40208

Xerces Society
Department of Zoology and Physiology
University of Wyoming
Laramie, WY 82071

The Amateur Entomological Society
Publications Agent
137 Gleneldon Road
Streatham, London SW 16, UK

Amateur Entomologist
109 Waveney Drive
Springfield
Chelmsford
Essex CM1 5QA, UK

E.W. Classey Ltd
Natural History Booksellers
P.O. Box 93
Faringdon
Oxon, UK

The Royal Entomological Society of London
41 Queen's Gate
London
SW7 5HU, UK
071–5848361

INDEX

Page numbers in *italic* refer to illustrations and captions.

ACKNOWLEDGMENTS

Quarto would like to thank the following people and organisations for permission to reproduce copyright material.

Key: bl = bottom left; br = bottom right; tr = top right; tl = top left; l = left; r = right; t = top; c = center; b = bottom

Premaphotos Wildlife
(c) **K.G. Preston-Mafham** 2; 4; 5tl, tr, cl, cr & br; 6; 9br; 11t & b; 19; 20t & b; 23; 26; 27; 31; 35t; 36; 41l; 53; 69; 70b; 71t & b; 72; 76l & r; 78l & r; 79; 81b; 83t, c & b; 84t; 89tl, tr & b; 90l & r; 95l; 97; 98r; 99; 105t & br; 107; 111t & b; 114b; 116tr; 118; 119t; 130r; 132; 134l; 135; 138b; 139t; 141l & r; 144l; 147t & b; 148; 149t & b, 152–3 (c) **R.A. Preston-Mafham** 38; 85; 98l; 101t & b; 116tl; 133b; 144r; 145 (c) **J. Preston-Mafham** 81tr

Wildlife Matters: John Feltwell
5bl; 10t & b; 22; 32; 37; 41r; 43; 46; 51; 52; 55; 58; 59; 61b; 62; 63; 64; 65; 75l & t; 86; 100t & b; 102; 113l & r; 114t; 115; 119b; 121; 123; 140t
Wildlife Matters: Michael Tweedie
7; 9t & bl; 25; 28; 33; 34; 35b; 39; 40; 50; 70t; 73; 81tl; 84b; 91t & b; 92t & b; 93; 95r; 96t & b; 100c; 103; 105bl; 109; 116b; 120; 123t, c & b; 124; 125; 128l & r; 129; 130l; 133t; 134r; 137; 138t; 139b; 140b; 146t, c & b; 150; 151t & b
Wildlife Matters: British Museum (Natural History)
21, 61t
Wildlife Matters: taken while on Project Wallace, courtesy of the Royal Entomological Society of London
49